T0222577

Practical Machine Learning with Rust

Creating Intelligent Applications in Rust

Joydeep Bhattacharjee

Apress®

Practical Machine Learning with Rust: Creating Intelligent Applications in Rust

Joydeep Bhattacharjee
Bangalore, India

ISBN-13 (pbk): 978-1-4842-5120-1   ISBN-13 (electronic): 978-1-4842-5121-8
https://doi.org/10.1007/978-1-4842-5121-8

Copyright © 2020 by Joydeep Bhattacharjee

Managing Director, Apress Media LLC: Welmoed Spahr
Acquisitions Editor: Celestin Suresh John
Development Editor: Matthew Moodie
Coordinating Editor: Aditee Mirashi

Cover designed by eStudioCalamar

Cover image designed by Freepik (www.freepik.com)

Distributed to the book trade worldwide by Springer Science+Business Media New York, 233 Spring Street, 6th Floor, New York, NY 10013. Phone 1-800-SPRINGER, fax (201) 348-4505, e-mail orders-ny@springer-sbm.com, or visit www.springeronline.com. Apress Media, LLC is a California LLC and the sole member (owner) is Springer Science + Business Media Finance Inc (SSBM Finance Inc). SSBM Finance Inc is a **Delaware** corporation.

For information on translations, please e-mail rights@apress.com, or visit http://www.apress.com/rights-permissions.

Apress titles may be purchased in bulk for academic, corporate, or promotional use. eBook versions and licenses are also available for most titles. For more information, reference our Print and eBook Bulk Sales web page at http://www.apress.com/bulk-sales.

Any source code or other supplementary material referenced by the author in this book is available to readers on GitHub via the book's product page, located at www.apress.com/978-1-4842-5120-1. For more detailed information, please visit http://www.apress.com/source-code.

Printed on acid-free paper

*To my wife, Saionee, for patiently hearing my
ideas and giving me advice, support, and motivation.
To my mom, father-in-law, and mother-in-law
for believing in me throughout the years.*

Table of Contents

About the Author

Joydeep Bhattacharjee is a Principal Engineer who works for Nineleaps Technology Solutions. After graduating from National Institute of Technology at Silchar, he started working in the software industry, where he stumbled upon Python. Through Python, he stumbled upon machine learning. He is the author of *fastText Quick Start Guide* (Packt, 2018). He has more than seven years' experience in the software industry and around four years developing machine learning applications. He finds great pleasure in developing intelligent systems that can parse and process data to solve challenging problems at work. He believes in sharing knowledge and loves mentoring. He also maintains a machine learning blog on Medium.

Acknowledgments

First and foremost, I would like to thank all the open source maintainers of the Rust crates mentioned in this book and the developers of the Rust languages itself, without which this book would not have been possible. Additionally, I would like to thank my friend Sherin Thomas for his help on the PyTorch sections.

Thanks to the Apress team for believing in me, to Celestin believing in my ideas, and to Aditee for pushing me on the initial drafts and for coordinating the whole process.

Introduction

This book is all about exploring Machine Learning in Rust lang. We will learn about the intricacies of creating machine learning applications and how they fit in the Rust worldview.

We will start from the very beginning by understanding some of the important concepts of Machine Learning as well as the basics of Rust lang. In the later chapters we will dive into the more specific areas of machine learning, such as data processing, computer vision, and natural language processing; and look at the Rust libraries that would make creating applications for those domains easier. We will also look at how to deploy those applications either onsite or over the cloud.

By the end of the book, the reader will have a solid understanding of creating high computation libraries using Rust. Armed with the knowledge of this amazing language, they can begin working toward creating applications that are more performant, memory safe, and less-resource heavy.

Who Is the Target Audience?

This book is best suited for the programmer who works in industrial optimization problems and is looking for ways to write better code and create better applications. Although this book does not assume any machine learning experience and will explain all concepts, it would still be best if there is some machine learning experience, especially using one of the major programming languages such as Python. This book does not assume any Rust knowledge and will be good for a budding Rust developer interested in machine learning or someone who is not satisfied with the current ecosystem and would like to take a look at the options available.

CHAPTER 1

Basics of Rust

In this chapter we will explore Rust as a language and the unique constructs and programming models that Rust supports. We will also discuss the biggest selling points of Rust and what makes this language particularly appealing to machine learning applications.

Once we have an overview of Rust as a language, we will start with its installation. Then we will move on to Cargo, which is the official package manager of Rust, and how we can create applications in Rust. Later we will look at Rust programming constructs such as variables, looping constructs, and the ownership model. We will end this chapter by writing unit tests for Rust code and showing how the tests can be run using the standard package manager.

By the end of this chapter, you should have a fair understanding of how to write and compile simple applications in Rust.

1.1 Why Rust?

There is a general understanding that there are differences between low-level systems programming languages and high-level application programming languages. Received wisdom says that if you want to create performant applications and create libraries that work on bare metal, you will need to work in a language such as C or C++, and if you want to create applications for business use cases, you need to program in languages such as Java/Python/JavaScript.

© Joydeep Bhattacharjee 2020
J. Bhattacharjee, *Practical Machine Learning with Rust*,
https://doi.org/10.1007/978-1-4842-5121-8_1

The aim of the Rust language is to sit in the intersection between high-level languages and low-level languages. Programs that are close to the metal necessarily handle memory directly, which means that there is no garbage collection in the mix. In high-level languages, memory is managed for the programmer.

Implementing garbage collection has costs associated with it. At the same time, garbage collection strategies that we have are not perfect, and there are still examples of memory leaks in programs with automatic memory management. One of the main reasons for memory leaks in higher-level languages are when packages are created to give an interface in the higher-level language but the core implementation is in a lower-level language. For example, the Python library pandas has quite a few memory leaks. Also, absence of evidence does not mean evidence of absence, and hence there is no formal proof that bringing in garbage collection strategies will remove all of the possible memory leaks.

The other issue is with referencing. References are easy to understand in principle, and they are inexpensive and essential to achieving developer performance in software creation. As such, languages that target low-level software creation such as C and C++ allow unrestricted creation of references and mutation of the referenced object.

1.2 A Better Reference

Typically, in object systems, objects live in a global object space called the heap or object store. There are no strict constraints on which part of the object store the object can access, because there are no restrictions on the way the object references are passed around. This has repercussions when preventing representation exposure for aggregate objects. The components that constitute an aggregate object are considered to be contained within that aggregate, and part of its representation. But, because the object store is global, there is, in general, no way to prevent other objects from accessing that representation. Enforcing the notion of containment with the standard reference semantics is impossible.

A better solution can be to restrict the visibility of different types of objects that are created. This is done by saying that all objects have a context associated with them. All paths from the root of the system must pass through the objects' owner.

In Rust, types maintain a couple of key invariants that are listed here. To start, every storage location is guaranteed to have either

- 1 mutable reference and 0 immutable references to it, or

- 0 mutable references and n immutable references to it.

We will see how this translates to actual code in a later part of this chapter. This invariant prevents races in concurrent systems as it prohibits concurrent reads and writes to a single memory location. By itself, however, this invariant is not enough to guarantee memory safety, especially in the presence of movable objects. For instance, since a given variable becomes unusable after the object has been moved, the storage locations associated with that variable may be reused or freed. As a result, any references previously created will be dangling after a move.

This issue is also resolved by the previous ownership rules in Rust. References to an object are created transitively through its owner. The type of system guarantees that the owner does not change after a move while references are outstanding. Conversely the type of systems allows change of ownership when there are no outstanding references. Examples of this will be discussed in more detail in a later part of the chapter.

All of what has just been mentioned is even more important in a machine learning application. Training machine learning applications involve a lot of data, and the more variation in the data the better, which translates to a lot of object creation during the training phase. You probably don't want memory errors. After deployment of the models, the models that get created are matrices and tensors, which are implemented as a collection of floats. It is probably not desirable to keep past objects and predictions dangling in memory even after they have no more use. There are advantages from creating a concurrent system as well. Rust types guarantee that there will be no race conditions and hence programmers can safely create machine learning applications that try to spread out the computation as much as possible.

Along with all this talk about memory safety and high performance, we also have high-level features such as type inference so we will not need to write types for all the variables. We will see when defining types are important in a later part of the chapter. Another interesting point from the earlier discussion is that when writing Rust code, we are not thinking about memory. So, from a usage point of view, it feels like memory is being managed for us. Then there are other constructs such as closures, iterators, and standard libraries, which make writing code in Rust more like writing in a high-level language. For machine learning applications, this is crucial. High-level languages such as Python have succeeded in machine learning because of the expressiveness of the language that supports free-form experimentation and developer productivity.

In this chapter we will be taking a look at the basics of Rust and the programming constructs that make Rust the language it is. We primarily cover topics such as Structs and Enums that look and feel different in this language and might not be what we would expect in this language. We will skip a lot of important things such as individual data types, which are similar to other languages such as C and Java. One of the core designs of Rust is that the programming feel should be the same as C and Java, which are more popular so that programmers coming from these languages don't have to do a lot of mental overhauling while also gaining a lot of memory advantages that have not been considered before.

1.3 Rust Installation

In this section we explore how to install Rust based on the operating system. The command and a possible output are shown.

```
$ curl https://sh.rustup.rs -sSf | sh

info: downloading installer

Welcome to Rust!

This will download and install the official compiler for the
Rust programming language, and its package manager, Cargo.

It will add the cargo, rustc, rustup and other commands to
Cargo's bin directory, located at:

 /home/ubuntu/.cargo/bin

This path will then be added to your PATH environment variable
by modifying the profile file located at:

 /home/ubuntu/.profile

You can uninstall at any time with rustup self uninstall and
these changes will be reverted.

Current installation options:

 default host triple: x86_64-unknown-linux-gnu
      default toolchain: stable
 modify PATH variable: yes

1) Proceed with installation (default)
2) Customize installation
3) Cancel installation
>1
```

```
info: syncing channel updates for 'stable-x86_64-unknown-linux-gnu'
info: latest update on 2019-02-28, rust version 1.33.0
(2aa4c46cf 2019-02-28)
info: downloading component 'rustc'
 84.7 MiB / 84.7 MiB (100 %) 67.4 MiB/s ETA: 0 s
info: downloading component 'rust-std'
 56.8 MiB / 56.8 MiB (100 %) 51.6 MiB/s ETA: 0 s
info: downloading component 'cargo'
info: downloading component 'rust-docs'
info: installing component 'rustc'
 84.7 MiB / 84.7 MiB (100 %) 10.8 MiB/s ETA: 0 s
info: installing component 'rust-std'
 56.8 MiB / 56.8 MiB (100 %) 12.6 MiB/s ETA: 0 s
info: installing component 'cargo'
info: installing component 'rust-docs'
 8.5 MiB / 8.5 MiB (100 %) 2.6 MiB/s ETA: 0 s
info: default toolchain set to 'stable'

 stable installed - rustc 1.33.0 (2aa4c46cf 2019-02-28)
```

Rust is installed now. Great!

To get started, you need Cargo's bin directory ($HOME/.cargo/
bin) in your PATHenvironment variable. Next time you log in
this will be done automatically.

To configure your current shell, run source $HOME/.cargo/env

If we study the earlier output, we will see the following points.

- rustup script has been successfully able to identify my
  distribution and will be installing rust binaries that are
  compatible with it.

- Installation of Rust along with Cargo (the official rust package manager) will be run through this command.

- The commands will be added to <home/>.cargo/bin and will be accessible from the command line.

1.4 Package Manager and Cargo

Cargo is a convenient build tool for development of Rust applications and libraries. The package information is supposed to be saved in a toml (Toms Obvious, Minimal Language) file. The toml file format is relatively new, and according to the github toml repo, it is designed to map unambiguously to a hash table.

1.5 Creating New Applications in Rust

Creating a new application is simple in Rust.

```
$ cargo new myfirstapp
         Created binary (application) `myfirstapp` package
```

Check the Cargo.toml file. You should see something like the following.

```
[package]
name = "myfirstapp"
version = "0.1.0"
authors = ["Joydeep Bhattacharjee"]
edition = "2018"

[dependencies]
```

As you can see, there is some basic information added here. Important items are the name and the version.

If you check the contents of the src/ folder, you can also see that there is a main.rs file. Check the contents of the main.rs file. You can see that there is a minimal file written with main function. To run a Rust app, you will need the main function that acts as the entry point for the code. The code in this case is a simple printing of hello world.

```
fn main() {
        println!("Hello, world!");
}
```

We can now build the application using the build command. This will generate a binary file that can be used to run the application. Once development of the application is done, we can use the --release flag to create an optimized binary. This needs to be done because by default, cargo builds disable many optimizations so that they are useful for testing. So when creating builds for production usage, the release flag should be used.

```
$ cargo build
   Compiling myfirstapp v0.1.0 (/tmp/myfirstapp)
        Finished dev [unoptimized + debuginfo] target(s) in 8.47s
$ ls target/debug/myfirstapp
target/debug/myfirstapp
$ ./target/debug/myfirstapp
Hello, world!
```

While developing, we can also use the cargo run command to shortcut the procedure just shown.

```
$ cargo run
        Finished dev [unoptimized + debuginfo] target(s) in 0.42s
        Running `target/debug/myfirstapp`
Hello, world!
```

1.6 Variables in Rust

In Rust, variables are defined using the let keyword. The types of the variables will be inferred for us. Take a look at the next example.

```
let x = "learning rust";
println!("{}", x);
```

println is used to see the variable.

There is a note on the special construct println! here. When you see the ! sign after a function name, that means that a macro is being used. Macros are special metaprogramming constructs in Rust, which are outside the scope of this book. The macro println is being used because Rust does not support variable args in functions and hence println has to be a macro.

We can see the type of the variable using the following code.

```
#![feature(core_intrinsics)]

fn print_type_of<T>(_: &T) {
        println!("{}", unsafe { std::intrinsics::
        type_name::<T>() });
}

fn main() {
        let x = "learning rust";

        println!("{}", x);
        print_type_of(&x);

}
```

This will not run in a default stable version though and will need to be compiled in the nightly version. The nightly compiler will need to be enabled. Nightly version is the place where unstable or potentially unstable code is kept, and so language features such as the one that we are discussing right now will only be available in a nightly version.

```
$ rustup default nightly
```

We should now be able to run the code.

```
$ ./variables1
learning rust
&str
```

Now try this out with different types of variables.

```
let x = "learning rust";
let y = 6;
let z = 3.14;

println!("{}", x);
println!("type of x:");
print_type_of(&x);
println!("type of y:");
print_type_of(&y);
println!("type of z:");
print_type_of(&z);
```

The output of using the above code is

```
$ ./variables1
learning rust
type of x:
&str
type of y:
i32
type of z:
f64
```

Note   i32 are essentially integers in 32 bit and f64 are floats in 64 bits. We will discuss different types of numbers throughout this book, but mostly Rust follows the primary data type formats that are universal in different languages for easy compilation into different architectures.

1.6.1 Mutation and Shadowing

The variables that are created using the let keyword are immutable. According to the Rust book, this is done because one of the primary focuses of Rust is safety.[1] When there is a need to change the values of variables, we can create variables that are mutable. This is done using the mut keyword with let.

```
let mut x = 32;
println!("Current value of x: {}", x);
x = 64;
println!("Current value of x: {}", x);
```

The output is

```
Current value of x: 32
Current value of x: 64
```

Mutating the type of the variable is not allowed though.

```
let mut x = 32;
println!("Current value of x: {}", x);
x = "rust";
println!("Current value of x: {}", x);
```

---

[1]https://doc.rust-lang.org/nightly/book/variable-bindings.html

So, for something like the previous example, we will get an error as shown.

```
$ rustc variables3.rs
error[E0308]: mismatched types
 --> variables3.rs:4:9
  |
4 |        x = "rust";
  |            ^^^^^^ expected integer, found reference
  |
  = note: expected type `{integer}`
          found type `&'static str`
```

error: aborting due to previous error

For more information about this error, try `rustc --explain E0308`.

Observe that in the place where x is assigned a string, the compiler is telling us that the code should have an integer. Passing a string now will not work.

For simple calculations we can use the shadowing principal as well. Shadowing happens when a variable declared within an outer scope is the same variable used within an inner scope. So something like what is shown here is perfectly valid.

```
fn main() {
    let x = 1;
    let x = x + 2;
    let x = x * 2;
    println!("Value of x: {}", x);
}
```

Output for the above code is Value of x: 6.

1.6.2 Variable Scoping

Also, in this case, the scope of the variables needs to be strictly maintained. Let's take a look at an example.

```
// variables5.rs
fn main() {
        let x = 5;

        if 4 < 10 {
                let x = 10;
                println!("Inside if x = {:?}", x);
        }
        println!("Outside if x = {:?}", x);
}
```

Check the output. Since the scope of the inner variable x ends after the first print statement, the first print statement prints x as 10 while the outer print statement prints 5.

```
$ ./variables5
Inside if x = 10
Outside if x = 5
```

As you can see, the scope of variables is maintained, and once the scope of the if statement is done, the variable x returns to the previous state.

1.7 Data Types

The data types are mostly similar to what you would find in other languages. Review the following list.

- bool : The Boolean type.

- char : A character type.

- i8 : The 8-bit signed integer type.

- i16 : The 16-bit signed integer type.

- i32 : The 32-bit signed integer type.

- i64 : The 64-bit signed integer type.

- isize : The pointer-sized signed integer type.

- u8 : The 8-bit unsigned integer type.

- u16 : The 16-bit unsigned integer type.

- u32 : The 32-bit unsigned integer type.

- u64 : The 64-bit unsigned integer type.

- usize : The pointer-sized unsigned integer type.

- f32 : The 32-bit floating-point type.

- f64 : The 64-bit floating-point type.

- array : A fixed-size array, denoted [T; N], for the element type, T, and the non-negative compile-time constant size, N.

- slice : A dynamically sized view into a contiguous sequence, [T].

- str : String slices.

- tuple : A finite heterogeneous sequence(T, U, . . .).

1.8 Functions

Functions are defined using the fn keyword. When defining functions, the signature of the function and the arguments will need to be said. We can skip the function signature for void functions. Remember that we did not provide any signature in case of the main function. To return from a function, there is no explicit return statement. Instead, return statements will not have the semicolon at the end.

```rust
fn main() {
    println!("{:?}", square_of(-5));
}

fn square_of(x: i32) -> i32 {
    println!("x = {:?}", x);
    x.pow(2)
}
// output:
//      x = -5
//      25
```

1.9 Conditions

There are two ways you can do conditional checking. One is the normal if blocks, and the other is pattern matching.

1.9.1 If Conditions

If conditions work similar to those in other languages. Only what you might not expect coming from more imperative languages is that an if block in Rust can return a statement. This is convenient as Rust has strict scoping rules as you saw before.

```rust
// Usage:
// $ ./conditions
// "cold"
fn main() {
    let place = "himalayas";
```

```
    let weather = if place == "himalayas" {
            "cold"
    } else {
            "hot"
    };
    println!("{:?}", weather);
}
```

Note   The println with a {:?} for showing the variable will work if
Debug has been implemented for the variable type. We will implement
the Debug for different classes in the later chapters, and this will be the
standard method of printing variables throughout this book.

1.9.2 Pattern Matching

For the same logic as shown in 1.9.1 we can use a more declarative syntax
using the match operator. The caveat is that all the use cases will need to
be covered. The above if can be written like what is shown next.

```
// Usage:
// $ ./match
// "cold"
fn main() {
    let place = "himalayas";

    let weather = match place {
            "himalayas" => "cold",
            _ => "hot",
    };
    println!("{:?}", weather);
}
```

1.10 References and Borrowing

As shown next, the functions can simply take a value, compute a result, and get the output.

```
// $ ./ownership1
// "rust 2018."

fn main() {
        let lang = "rust";
        let rust1 = add_version(&lang);
        println!("{:?}", rust1);
}

fn add_version(s: &str) -> String {
        s.to_string() + " 2018."
}
```

But this simple logic is not necessarily the case when writing real-world apps. Variables are generally created so that they can be used in multiple places. Let's take the next example, where we are trying to use the same variable lang in different functions.

```
// ownership2.rs
fn main() {
    let lang = String::from("rust");
    let rust1 = add_version(lang);
    println!("{:?}", rust1);
    let rust2 = add_lang(lang);
    println!("{:?}", rust2);
}

fn add_version(s: String) -> String {
    s + " " + "2018!!"
}
```

```
fn add_lang(s: String) -> String {
    s + " " + "lang."
}
```

Compiling this code gives the following error.

```
$ rustc ownership2_invalid.rs
error[E0382]: use of moved value: `lang`
 --> ownership2_invalid.rs:8:23
   |
6  |        let rust1 = add_version(lang);
   |                                ---- value moved here
7  |        println!("{:?}", rust1);
8  |        let rust2 = add_lang(lang);
   |                             ^^^^ value used here after move
   |
   = note: move occurs because `lang` has type
   `std::string::String`, which does not implement the `Copy` trait

error: aborting due to previous error

For more information about this error, try `rustc --explain E0382`.
```

So here comes the concept of borrowership. Using the same variable multiple times is not considered safe, and so what needs to be done is to create a reference to the variable and pass the reference around. This is similar to the analogy of borrowing books. The same book can be borrowed by multiple people, but it can be owned by only one person. Referencing is done using the ampersand & operator. The earlier code will be changed to the following. In this case, sadly, we lose our implementation of add and will need to use a method.

```rust
// ownership3.rs

fn main() {
        let lang = String::from("rust");
        let rust1 = add_version(&lang);
        println!("{:?}", rust1);
        let rust2 = add_lang(&lang);
        println!("{:?}", rust2);
}

fn add_version(s: &String) -> String {
        s.push_str(" 2019!!");
        s.to_string()
}

fn add_lang(s: &String) -> String {
        s.push_str(" lang.");
        s.to_string()
}
```

Now if we run this code, we get the following error. Notice that the error is different. If you are from a C++ background, you might notice that this is a common construct. But in Rust, trying to push a string to a vector seems to through an error.

```
error[E0596]: cannot borrow immutable borrowed content `*s` as
mutable
  --> ownership2_invalid2.rs:12:2
   |
11 |  fn add_version(s: &String) -> String {
   |                    ------- use `&mut String` here to make
                            mutable
12 |      s.push_str(" 2019!!");
   |      ^ cannot borrow as mutable
```

```
error[E0596]: cannot borrow immutable borrowed content `*s` as
mutable
  --> ownership2_invalid2.rs:17:2
   |
16 | fn add_lang(s: &String) -> String {
   |                ------- use `&mut String` here to make
                            mutable
17 |     s.push_str(" lang.");
   |     ^ cannot borrow as mutable
```

error: aborting due to 2 previous errors

For more information about this error, try `rustc --explain
E0596`.

So the error is that you cannot borrow a reference and try to mutate
it. This is one of the defining characteristics of Rust as discussed in the
first section, and we will need to keep this in mind all the time later in the
development process.

1.10.1 Mutable References

As we have seen in the variable section, the let keyword essentially creates
variables that are immutable. A way to convert them to mutable variables
is by putting in the mut keyword. So, in the previous code, put all the places
where the variable is referenced as mutable. The resulting code is shown
here.

```
// $ ./ownership4
// "rust 2019!!"
// "rust 2019!! lang."
```

```
fn main() {
        let mut lang = String::from("rust");
        let rust1 = add_version(&mut lang);
        println!("{:?}", rust1);
        let rust2 = add_lang(&mut lang);
        println!("{:?}", rust2);
}

fn add_version(s: &mut String) -> String {
        s.push_str(" 2019!!");
        s.to_string()
}

fn add_lang(s: &mut String) -> String {
        s.push_str(" lang.");
        s.to_string()
}
```

We have now achieved our objective to be able to pass the variables to all the functions. There is a final problem though. As the initial string variable is mutable, the same object is getting changed in all the functions. As you can see, the output is "rust 2019!!" and in the next function the "lang" gets added to "rust 2019" and not "rust" alone. This can be remedied by using str, which is the immutable version of the String type. The code changes are shown here with comments by the code changes.

```
// $ ./ownership
// "rust 2018."
// "rust lang."

fn main() {
        let lang = "rust"; // major change
        let rust1 = add_version(&lang);
        println!("{:?}", rust1);
```

```
        let rust2 = add_lang(&lang);
        println!("{:?}", rust2);
}

fn add_version(s: &str) -> String {
        s.to_string() + " 2018." // major change
}

fn add_lang(s: &str) -> String {
        s.to_string() + " lang." // major change
}
```

1.11 Object-Oriented Programming

Rust has a healthy dose of both functional programming paradigms as well
as the OOP (Object-Oriented Programming) paradigm, but since the readers
of this book are probably coming from a machine learning industry and are
interested in learning machine learning (that is the assumption that we are
making), and because object-based programming models are in vogue in
these fields (as they are in almost all the programming fields), we will take a
special look at object-based models of programming. To get into this, we will
start with the core types and then go to how things come together in Rust.

1.11.1 Structures

First come structures. Structures are essentially named tuples, and
they can be utilized to store organized information. They can also be
considered as attributes of an instance. Take a look at the next code.

```
// structures.rs
// $ ./structures
// Planet { co2: 0.04, nitrogen: 78.09 }
// Planet { co2: 95.32, nitrogen: 2.7 }
```

```
#[derive(Debug)]
struct Planet {
        co2: f32,
        nitrogen: f32
}

fn main() {
        let earth = Planet { co2: 0.04, nitrogen: 78.09 };
        println!("{:?}", earth);

        let mars = Planet { co2: 95.32, nitrogen: 2.7 };
        println!("{:?}", mars);
}
```

Also notice that when creating the struct, we have annotated with the Debug trait. This is done because it lets us print the class instances earth and mars in the main method in the debug mode. Because of this we will get output something like what comes next, clearly showing the class and the variable names and values.

```
// $ ./oops
// Planet { co2: 0.04, nitrogen: 78.09 }
// Planet { co2: 95.32, nitrogen: 2.7 }
```

1.11.2 Traits

Structs and other data structures do not have or own functionality. These functionalities are defined using traits. Traits are similar to interfaces in OOP languages. One of the core principles of Rust is the principle of zero-cost abstraction. Zero-cost abstraction in the words of Bjarne Stroustrup:

> *C++ implementations obey the zero-overhead principle: What you don't use, you don't pay for [Stroustrup, 1994]. And further: What you do use, you couldn't hand code any better.*
>
> —Stroustrup

Traits are meant to support zero-cost abstractions. A trait can be implemented for multiple types and new functionality can be implemented for old types using traits. An example of a simple trait is shown next. co2 and nitrogen are members of the trait, whereas summarize is a method.

```
trait Atmosphere {
        fn new(co2: f32, nitrogen: f32) -> Self;
        fn amount_of_other_gases(&self) -> f32;
        fn summarize(&self);
}
```

1.11.3 Methods and impl

You can implement methods on structs using the impl block. In this way we have something that resembles objects, which means that we have a type that encapsulates data and behavior. An example combining the previous struct and trait are shown next as well as how the implementation is done for the specific struct.

```
// oops.rs

#[derive(Debug)]
struct Planet {
    co2: f32,
    nitrogen: f32
}

trait Atmosphere {
        fn new(co2: f32, nitrogen: f32) -> Self;
        fn amount_of_other_gases(&self) -> f32;
        fn summarize(&self);
}
```

```rust
impl Atmosphere for Planet {
        fn new(co2: f32, nitrogen: f32) -> Planet {
                Planet { co2: co2, nitrogen: nitrogen }
        }

        fn amount_of_other_gases(&self) -> f32 {
                100.0 - self.co2 - self.nitrogen
        }

        fn summarize(&self) {
                let other_gases = self.amount_of_other_gases();
                println!("For planet {planet:?}: co2 = {co2},
                nitrogen={nitrogen}, other_gases={other_gases}",
                        planet=self, co2=self.co2,
                        nitrogen=self.nitrogen,
                        other_gases=other_gases);
        }
}

fn main() {
    let earth = Planet { co2: 0.04, nitrogen: 78.09 };
    println!("{:?}", earth);

    let mars = Planet { co2: 95.32, nitrogen: 2.7 };
    println!("{:?}", mars);

    earth.summarize();

    mars.summarize();
}
```

For the previous code, we will see the output next. You can see that after the creation of the instances earth and mars, the summarize method for the instances run, which makes the calculation for the different parameters and prints the result.

```
// $ ./oops
// Planet { co2: 0.04, nitrogen: 78.09 }
// Planet { co2: 95.32, nitrogen: 2.7 }
// For planet Planet { co2: 0.04, nitrogen: 78.09 }:
   co2 = 0.04, nitrogen=78.09, other_gases=21.870003
// For planet Planet { co2: 95.32, nitrogen: 2.7 }:
   co2 = 95.32, nitrogen=2.7, other_gases=1.9800003
```

1.11.4 Enumerations

Enumerations are an interesting way to create a type that will hold one of
the values of few different variants. For example, IP addresses can have
one of the types of IP addresses but not both. A book can be hardbound or
paperback but not both and so on. In such cases, we can define the types
as enums. Take a look at the following code.

```rust
// enumerations.rs
#[derive(Debug)]
enum NationalHolidays {
      GandhiJayanti,
      RepublicDay,
      IndependenceDay,
}

fn inspect(day: NationalHolidays) -> String {
      match day {
      NationalHolidays::GandhiJayanti => String::from("Oct 2"),
      NationalHolidays::RepublicDay => String::from("Jan 26"),
      NationalHolidays::IndependenceDay => String::from
      ("Aug 15"),
      }
}
```

```rust
fn main() {
    let day = NationalHolidays::GandhiJayanti;
    let date = inspect(day);
    println!("{:?}", date); // output: Oct 2
}
```

Similar to structs, we can attach functionality over enums as well. Running this is left as an exercise for the reader.

1.12 Writing Tests

Finally, as part of the core basics that we are covering in Rust, we are at the place of how to write tests in Rust. As an example, we are taking the ownership1.rs code and writing a test for the add_version function. The code for the method is simple with a string "2018" concatenated to the input string.

To write the test for this method, we will need to write a function and annotate the function with #[test]. Then the compiler would know that the underlying function is a test method. The assert_eq function in Rust is a macro. After that, the first is the actual output from the function, which in this case is add_version function and the next parameter is the expected value, which is String::from("abcd 2018.").

```rust
fn add_version(s: &str) -> String {
    s.to_string() + " 2018."
}

#[test]
fn test_add_version() {
    assert_eq!(add_version("abcd"), String::from("abcd
    2018."));
}
```

If written in a Cargo package, we can use the `cargo test` command to compile and run the test for us.

```
$ cargo test
    Compiling unittestingexample v0.1.0 (/path/to/code)
        Finished dev [unoptimized + debuginfo] target(s) in 9.75s
        Running target/debug/deps/unittestingexample-
        9d94da4358544e8a

running 1 test
test test_add_version ... ok

test result: ok. 1 passed; 0 failed; 0 ignored; 0 measured; 0
filtered out
```

Note that a new test binary gets created, and the test binary is run to evaluate the test cases.

1.13 Summary

This chapter introduces you to the basics of Rust by showing how to install Rust and set up a development environment, how to start working on Rust projects using Cargo, and how to use variables and functions in Rust. You also had a glimpse of ownership and references in Rust and how to implement some object paradigms using Structs, Enums, Traits, and Impl's.

In the next chapter, you will look at using Rust and creating Rust programs that use machine learning algorithms such as linear or logistic regression.

1.14 References

[1] Pavel Kordík. On Machine Learning and
    Programming Languages. Ed. by Towards Data
    Science. [Online; accessed 23-May-2019]. 2018.
    URL: https://medium.com/recombee-blog/
    machinelearning-for-recommender-systems-
    part-1-algorithmsevaluation-and-cold-start-
    6f696683d0ed

[2] PyTorch bindings for Rust and OCaml. [Online;
    accessed 11-Nov-2019]. 2019. URL: https://
    www.reddit.com/r/MachineLearning/comments/
    axy689/p_pytorch_bindings_for_rust_and_
    ocaml/

[3] Oxide: The Essence of Rust. by Aaron Weiss, Daniel
    Patterson, Nicholas D. Matsakis, Amal Ahmed.
    [Online; accessed 11-Nov-2019]. 2019. URL:
    https://arxiv.org/abs/1903.00982

[4] Short Paper: Rusty Types for Solid Safety. by Sergio
    Benitez [Online; accessed 11-Nov-2019]. 2019. URL:
    https://sergio.bz/docs/rusty-types-2016.pdf

[5] Ownership Types for Flexible Alias Protection
    by David G. Clarke, John M. Potter, James Noble
    [Online; accessed 11-Nov-2019]. 2019. URL: http://
    citeseerx.ist.psu.edu/viewdoc/download;?
    doi=10.1.1.23.2115&rep=rep1&type=pdf

[6]   Java Garbage Collection handbook [Online;
       accessed 11-Nov-2019]. 2019. URL: `https://`
       `plumbr.io/java-garbage-collection-handbook`

[7]   toml github repo [Online; accessed 11-Nov-2019].
       2019. URL: `https://github.com/toml-lang/toml`

[8]   Rust: str vs String [Online; accessed 12-Oct-2019].
       2017. URL: `https://www.ameyalokare.com/`
       `rust/2017/10/12/rust-str-vs-String.html`

CHAPTER 2

Supervised Learning

2.1 What Is Machine Learning?

Machine learning is the science of getting computers to act without being specifically programmed. This is done by implementing special algorithms that have the ability to detect patterns in data. From a developer point of view, this means creating a system that has access to relevant data, is able to feed the data to machine learning algorithms, and is able to take the output and redirect it to downstream processes and tasks.

Supervised learning Supervised learning happens when you pass both the input and the desired outputs to the system, and you want the resulting machine learning model to capture the relationship. Supervised learning is again divided into two subsections based on the type of the labels.

Supervised tasks when the target variable is continuous are termed as regression problems. For example, the price of a product can be any value. We should probably go for regression techniques when the prediction needs to be made on something that requires the prediction of a quantity of some sort. Quality of a regression model is generally measured using some form of error measures, that is, the difference between the target values and the predicted values.

Classification problems are different from regression problems in that the labels are discrete and finite. For example, we can categorize an email message as spam or ham. So, a problem is a classification problem when we are interested in the resulting bucket that a particular set of

© Joydeep Bhattacharjee 2020
J. Bhattacharjee, *Practical Machine Learning with Rust*,
https://doi.org/10.1007/978-1-4842-5121-8_2

feature readings will fall into. Quality of a classification model is generally measured using an accuracy measure that is essentially the count of the number of times the model has been right.

Unsupervised learning In unsupervised algorithms, the labels or the target classes are not given. So, the goal of unsupervised learning is to attempt to find natural partitions of patterns.

The best time for unsupervised learning is when the data on desired outcomes is not known, such as creating a new class of products that the business has never sold before.

Reinforcement learning In reinforcement learning, specifically crafted agents are released in an environment, in which they learn to take actions based on some notion of rewards and punishments. Reinforcement learning has been applied to robotic movements and different classes of games with some success.

In this chapter we will be taking a look at creating models for different machine learning algorithms using Rust. We will first read a dataset from a csv file. This is a common dataset and will be representative of the types of data in the real world. Then we will look at the logic of popular algorithms, why they work, and how to implement them using some Rust machine learning packages such as `rusty_machine`. We will also take a look at how to evaluate the accuracy of each model.

By the end of this chapter, you should have a fair understanding of how to create common machine learning models and implement them in Rust.

2.2 Dataset Specific Code

Before going into regression algorithms and the associated code, let's build the surrounding boilerplate code. Look at the *rusty_machine_regression* package in the code that is shared with this book.

For showing usage of regression, we will be using the boston housing dataset.[1] You can also download the data from the kaggle page for boston dataset.[2] The dataset has 14 features and has 506 samples. The following is a description of the dataset.

- CRIM - per capita crime rate by town
- ZN - proportion of residential land zoned for lots over 25,000 sq. ft.
- INDUS - proportion of non-retail business acres per town
- CHAS - Charles River dummy variable (1 if tract bounds river; 0 otherwise)
- NOX - nitric oxides concentration (parts per 10 million)
- RM - average number of rooms per dwelling
- AGE - proportion of owner-occupied units built prior to 1940
- DIS - weighted distances to five Boston employment centers
- RAD - index of accessibility to radial highways
- TAX - full-value property-tax rate per $10,000
- PTRATIO - pupil-teacher ratio by town
- B - 1000(Bk - 0.63)$\hat{2}$ where Bk is the proportion of blacks by town
- LSTAT - % lower status of the population
- MEDV - Median value of owner-occupied homes in $1000's

[1]https://www.cs.toronto.edu/ delve/data/boston/bostonDetail.html.
[2]https://www.kaggle.com/c/boston-housing/data.

Now let's first create the project. We can create a project with the cargo command as shown next in Listing 2-1.

Listing 2-1. Create new package

```
$ cargo new rustymachine-regression --bin
$ cd rustymachine-regression
```

We can see that an src/main.rs file has been created and a Cargo.toml file has been created in the directory. We can create additional directory data using the command mkdir data to keep the boston housing dataset there.

The data is kept in a file "data/housing.csv". A quick look at the csv suggests that it is a fixed width file with no headers (Listing 2-2).

Listing 2-2. Housing data

```
$ head -n1 data/housing.csv \
    | grep -Eo '[0-9]*[.]?[0-9]*' \
    | wc -l
14
$ head -n2 data/housing.csv
 0.00632  18.00   ...    4.98  24.00
 0.02731   0.00   ...    9.14  21.60
```

To parse through the files and create our model, we will need to depend on various external dependencies. Rust packages are generally called crates. To access the code in the different crates, we will need to add them in the Cargo.toml file. The crates csv parses the csv files with the help of additional crates serde and serde-derive. We will talk about the rusty-machine crate later so for now just add them to the toml file. Notice that we are using Rust version 2018. This is the version that we will be using throughout the book (Listing 2-3).

Listing 2-3. chapter2/rustymachine_regression/Cargo.toml

```
[package]
name = "rustymachine_regression"
version = "0.1.0"
edition = "2018"

[dependencies]
rusty-machine = "0.5.4"
serde = "1"
serde_derive = "1"
rand = "0.6.5"
csv = "1.0.7"
```

For your applications, go to the `crates.io` page to find the available versions of the respective dependencies for your applications. In the website you can see that there is a search bar where you can put your search terms. In Figure 2-1, the `csv` page is shown.

If you scroll below, you will get the accompanying documentation as well.

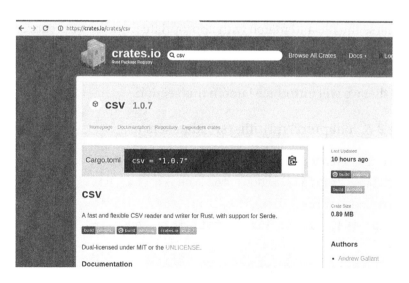

Figure 2-1. *Crates page*

Now we will start writing the code for the first module in this book. Most of the code in this book would be written in the src/main.rs file unless otherwise stated. Also, you can understand the file where the code is written in the label for the code block. We will go ahead and add required modules that will be imported using the use statement (Listing 2-4).

Listing 2-4. chapter2/rustymachine_regression/src/lin_reg.rs

```
extern crate serde;
#[macro_use]
extern crate serde_derive;
use std::io::prelude::*;
use std::io::BufReader;
use std::path::Path;
use std::fs::File;
use std::vec::Vec;
use std::error::Error;
use rand::thread_rng;
use rand::seq::SliceRandom;
```

To parse the file and match with appropriate records, we create a struct with each value to be a float64 (Listing 2-5).[3] We will need to initialize them as float64 due to the Rust machine learning library to we will use that we will introduce later in this section.

Listing 2-5. chapter2/ml-utils/src/datasets.rs

```
pub struct BostonHousing {
  crim: f64, zn: f64, indus: f64, chas: f64, nox: f64,
  rm: f64, age: f64, dis: f64, rad: f64, tax: f64,
  ptratio: f64, black: f64, lstat: f64, medv: f64,
}
```

---

[3]This does not mean that they cannot be used in 32-bit machines. Floats have nothing to do with the bitness of the machine.

One way to convert the file into records is to read the file line by line and build the BostonHousing record. This should enable us to implement methods that separate out the predictors and the responses or target variables. In this case we consider the median value of the house (the price of the house) as the target, and we will try to predict the price of the house based on the predictors available.

Listing 2-6 shows the code to implement the building of a BostonHousing record and methods to build the feature vector and targets.

Listing 2-6. chapter2/ml-utils/src/datasets.rs

```
impl BostonHousing {
  pub fn new(v: Vec<&str>) -> BostonHousing {
    let f64_formatted: Vec<f64> = v.iter().map(
      |s| s.parse().unwrap()).collect();
    BostonHousing {
          crim: f64_formatted[0], zn: f64_formatted[1],
          indus: f64_formatted[2], chas: f64_formatted[3],
          nox: f64_formatted[4], rm: f64_formatted[5],
          age: f64_formatted[6], dis: f64_formatted[7],
          rad: f64_formatted[8], tax: f64_formatted[9],
          ptratio: f64_formatted[10], black: f64_formatted[11],
          lstat: f64_formatted[12], medv: f64_formatted[13] }
  }

  pub fn into_feature_vector(&self) -> Vec<f64> {
    vec![self.crim, self.zn, self.indus, self.chas, self.nox,
      self.rm, self.age, self.dis, self.rad,
      self.tax, self.ptratio, self.black, self.lstat]
  }
```

```
  pub fn into_targets(&self) -> f64 {
    self.medv
  }
}
```

Now that we can proceed and start reading the file line by line using get_boston_records_from_file file. Each line will be passed to the get_boston_record function, which splits the string to a vector of strings to be parsed appropriately by BostonRecord (Listing 2-7).

Listing 2-7. chapter2/ml-utils/src/datasets.rs

```
fn get_boston_record(s: String) -> BostonHousing {
  let v: Vec<&str> = s.split_whitespace().collect();
  let b: BostonHousing = BostonHousing::new(v);
  b
}

fn get_boston_records_from_file(
    fl: impl AsRef<Path>) -> Vec<BostonHousing> {
  let file = File::open(fl).expect("no such file");
  let buf = BufReader::new(file);
  buf.lines().enumerate()
    .map(|(n, 1)| 1.expect(
      &format!("Could not parse line no {}", n)))
    .map(|r| get_boston_record(r))
    .collect()
}
```

The output of get_boston_records_from_file will now be stored in a mutable data variable to be used later (Listing 2-8).

Note Since these functions such as get_boston_record and get_boston_records_from_file and other dataset parsing functions are kept in the ml-utils package that is created in the chapter2 directory, a separate package has been created for these functions as we will be using these datasets for various machine learning models throughout the book. The code for the models will be kept in the respective chapter and packages. Hence, we will also see the path to ml-utils being referenced to in the cargo file and the module importing part of the code. Please have a look at the code shared with the book for a better understanding.

Listing 2-8. chapter2/rustymachine_regression/src/lin_reg.rs

```
use ml_utils::datasets::get_boston_records_from_file;

pub fn run() -> Result<(), Box<dyn Error>> {
  let fl = "data/housing.csv";
  let mut data = get_boston_records_from_file(&fl);
  // remaining part of the function
```

In machine learning tasks, it is a good practice to shuffle the incoming dataset. Shuffling data serves the purpose of reducing variance and making sure that models remain general and overfit less. Shuffling helps in making sure that the train/test/validation samples of the dataset are representative of the overall distribution of the data.

In Listing 2-9 we shuffle the data, split them into 80% train and 20% tests, and convert them into f64 vectors.

Listing 2-9. chapter2/rustymachine_regression/src/lin_reg.rs

```
pub fn run() -> Result<(), Box<dyn Error>> {
  // previous code

  data.shuffle(&mut thread_rng());

  // separate out to train and test datasets.
  let test_size: f64 = 0.2;
  let test_size: f64 = data.len() as f64 * test_size;
  let test_size = test_size.round() as usize;
  let (test_data, train_data) = data.split_at(test_size);
  let train_size = train_data.len();
  let test_size = test_data.len();

  // differentiate the predictors and the targets.
  let boston_x_train: Vec<f64> = train_data.iter()
    .flat_map(|r| r.into_feature_vector())
    .collect();
  let boston_y_train: Vec<f64> = train_data.iter()
    .map(|r| r.into_targets()).collect();
  let boston_x_test: Vec<f64> = test_data.iter()
    .flat_map(|r| r.into_feature_vector()).collect();
  let boston_y_test: Vec<f64> = test_data.iter()
    .map(|r| r.into_targets()).collect();

  // remaining part of the function
```

Data has been loaded and split into train and test components. The code in Listing 2-9 is generic and will be there in most of the run methods used in the run methods of the regression models in this chapter.

2.3 Rusty_Machine Library

Rusty_machine is a general-purpose machine learning library written entirely in Rust. The main aims of the rusty_machine are ease of use and speed without having to depend on a huge number of external libraries. The consistency in the api is achieved using the rust's trait system. It currently uses rulinalg for its linear algebra back end. In this book, one of the core libraries that we will focus on to achieve our machine learning needs is by using this library.

To use the rusty machine library, we will need to convert the data vectors into rulinalg supported Matrices. Convert the above vectors like those shown in Listing 2-10.

Listing 2-10. chapter2/rustymachine_regression/src/lin_reg.rs

```
use rusty_machine;
use rusty_machine::linalg::Matrix;
use rusty_machine::linalg::Vector;

pub fn run() -> Result<(), Box<dyn Error>> {
  // previous code ...

  let boston_x_train = Matrix::new(train_size, 13,
  boston_x_train);
  let boston_y_train = Vector::new(boston_y_train);
  let boston_x_test = Matrix::new(test_size, 13, boston_x_test);
  let boston_y_test = Matrix::new(test_size, 1, boston_y_test);

  // remaining part ...
```

In the code shown in Listing 2-10, we have boston_y_train as a vector but boston_y_test as a matrix of dimension 1. Theoretically they are the same but the datatypes have been taken as different because later on we will be using the function neg_mean_squared_error, which expects the inputs to be in matrix format as seen in Listing 2-11. Since we have not reached that stage yet, you can choose to keep boston_y_test as a vector for now.

Listing 2-11. Error in lin_reg.rs

```
error[E0308]: mismatched types
  --> src/lin_reg.rs:79:52
   |
79 |       let acc = neg_mean_squared_error(&predictions,
         &boston_y_test);
   |                                                  ^^^^^^^^^^^^^^
   expected struct `rulinalg::matrix::Matrix`, found struct
   `rulinalg::vector::Vector`
   |
   = note: expected type `&rulinalg::matrix::Matrix<f64>`
             found type `&rulinalg::vector::Vector<f64>`

error: aborting due to previous error
```

In rusty machine the two key methods that are implemented throughout all the model classes are the train and predict methods. Readers familiar with the sciki-learn api might be comfortable in it. As the name suggests, we will need to pass the training data to the train method for the respective models and the test data to the predict method of the respective models.

2.4 Linear Regression

Ordinary Least Squares linear regression is the method where a linear model with coefficients $w = (w1,w2,...,wp)$ is optimized to minimize the residual sum of squares between the observed responses in the dataset, and the responses predicted by the linear approximation. Mathematically it solves the problem of the form:

$$\min_{\omega} = \left\| X\omega - y \right\|^2 \tag{1}$$

Using Rusty_Machine One of the methods to optimize the model is Gradient Descent. Train a linear regression model using the code in Listing 2-12.

Listing 2-12. chapter2/rustymachine_regression/src/lin_reg.rs

```
pub fn run() -> Result<(), Box<dyn Error>> {
  // previous code...

  let mut lin_model = LinRegressor::default();
  lin_model.train(&boston_x_train, &boston_y_train)?;

  let predictions = lin_model.predict(&boston_x_test).unwrap();
  let predictions = Matrix::new(test_size, 1, predictions);
  let acc = neg_mean_squared_error(&predictions, &boston_y_test);
  println!("linear regression error: {:?}", acc);
  println!("linear regression R2 score: {:?}", r_squared_score(
    &boston_y_test.data(), &predictions.data()));

  Ok(())
}
```

The output for the code in Listing 2-12 should be similar to that in Listing 2-13.

Listing 2-13. lin_reg.rs output

```
$ cargo run lr < ../datasets/housing.csv
LinRegressor { parameters: None }
linear regression error: -23.242663146907553
linear regression R2 score: 0.7627858374713654
```

Creating a linear regression model similar to Listing 2-13 is kept in the module chapter2/rustymachine_regression/src/lin_reg.rs module.

Using Tensorflow An interesting implementation of regression can be done using the closed form solution of the regression equation using Tensorflow. In this section we will implement the closed form solution using Rust implementation of Tensorflow.

Tensorflow is designed in such a way that to implement computations, we essentially need to create graphs. A Tensorflow program is essentially divided into two parts. In one part we create the computational graph and then we write the code that runs the graph in a session. It solves linear regression using the normal equation where we solve for theta $\left(\hat{\theta} = \left(X^T \cdot X\right)^{-1} \cdot X^T \cdot y\right)$ where $\hat{\theta}$ are the weights.

To create the computational graph, we will first load the vectors to the tensors with the appropriate dimensions (Listing 2-14).

Listing 2-14. chapter/rust_and_tf/src/linear_regression.rs

```
pub fn run() -> Result<(), Box<dyn Error>> {
  // previous code ...

  let dim = (boston_y_train.len() as u64, 13);
  let test_dim = (boston_y_test.len() as u64, dim.1);
  let X_train = <Tensor<f64>>::new(&[dim.0, dim.1])
    .with_values(&boston_x_train)?;
  let y_train = <Tensor<f64>>::new(&[dim.0, 1])
    .with_values(&boston_y_train)?;
  let X_test = <Tensor<f64>>::new(&[test_dim.0,
                                   test_dim.1])
    .with_values(&boston_x_test)?;

  // rest of the code ...
```

Note that in the previous listing the starting of the function is similar to what we have seen in Linear Regression sections Listings 2-9 and earlier.

We will need to also calculate the transpose as you can see from the closed form solution. Transposing using Tensorflow is not that easy and hence we can take help from the transpose library with the version being `transpose = "0.2.0"`. The whole cargo dependencies is shown in Listing 2-15. The crates serde and serde-derive is used for data serialisation and deserialisation; rand crate is for generating random numbers, mnist crate for easy access of the mnist dataset, random crate provides some good random functions which we will use in the code, mnist is the crate that we will developed and is available in book code. Important crates in this section is the transpose crate which we will use for transposing the matrices and the tensorflow crate for accessing the tensorflow functions.

Listing 2-15. chapter2/rust_and_tf/Cargo.toml

```
[package]
name = "rust_and_tf"
version = "0.1.0"
edition = "2018"

[dependencies]
tensorflow = { version = "0.13.0", features = ["tensorflow_
unstable"] }
serde = "1"
serde_derive = "1"
rand = "0.6.5"
transpose = "0.2.0"
mnist = "0.4.0"
ml-utils = { path = "../ml-utils" }
random = "0.12.2"
```

And then use the transpose crate to transpose `boston_x_train` (Listing 2-16).

Listing 2-16. chapter2/rust_and_tf/src/linear_regression.rs

```
pub fn run() -> Result<(), Box<dyn Error>> {
  // previous code ...

  let mut output_array = vec![0.0; (dim.0 * dim.1) as usize];
  transpose::transpose(&boston_x_train,
    &mut output_array,
    dim.1 as usize, dim.0 as usize);
  let XT =  <Tensor<f64>>::new(&[dim.1, dim.0]).with_
  values(&output_array[..])?;

  // remaining code ...
```

We should now be able to create the graph (Listing 2-17). A tensorflow graph represents the data flow of the computations. We can code the specific computations that will go as part of the graph as shown in Listing 2-17.

Listing 2-17. chapter2/rust_and_tf/src/linear_regression.rs

```
pub fn run() -> Result<(), Box<dyn Error>> {
  // previous code ...
  let mut graph = Graph::new();

  let XT_const = {
    let mut op = graph.new_operation("Const", "XT")?;
    op.set_attr_tensor("value", XT)?;
    op.set_attr_type("dtype", DataType::Double)?;
    op.finish()?
  };

  let X_const = {
    let mut op = graph.new_operation("Const", "X_train")?;
    op.set_attr_tensor("value", X_train)?;
    op.set_attr_type("dtype", DataType::Double)?;
```

```
  op.finish()?
};
let y_const = {
  let mut op = graph.new_operation("Const", "y_train")?;
  op.set_attr_tensor("value", y_train)?;
  op.set_attr_type("dtype", DataType::Double)?;
  op.finish()?
};
let mul = {
  let mut op = graph.new_operation("MatMul", "mul")?;
  op.add_input(XT_const.clone());
  op.add_input(X_const);
  op.finish()?
};
let inverse = {
  let mut op = graph.new_operation("MatrixInverse", "mul_inv")?;
  op.add_input(mul);
  op.finish()?
};
let mul2 = {
  let mut op = graph.new_operation("MatMul", "mul2")?;
  op.add_input(inverse);
  op.add_input(XT_const.clone());
  op.finish()?
};
let theta = {
  let mut op = graph.new_operation("MatMul", "theta")?;
  op.add_input(mul2);
  op.add_input(y_const);
op.finish()?
};

// remaining code ...
```

Notice that in the code in Listing 2-17, different operations are defined by the new_operation method on the graph. The inputs are defined and any other related attributes may be defined. Essentially, we are calling the C++ api directly in the code. To get a list of the operations defined in the api, take a look at the array-ops page of Tensorflow at https://www.tensorflow.org/api_docs/cc/group/array-ops.

This is only part of the graph. In the above graph the equation has been defined and the θ values are being computed. Now we find the predicted values as well as shown in Listing 2-18.

Listing 2-18. chapter2/rust_and_tf/src/linear_regression.rs

```
pub fn run() -> Result<(), Box<dyn Error>> {
  // previous code ...

  let X_test_const = {
    let mut op = graph.new_operation("Const", "X_test")?;
    op.set_attr_tensor("value", X_test)?;
    op.set_attr_type("dtype", DataType::Double)?;
    op.finish()?
  };
  let predictions = {
    let mut op = graph.new_operation("MatMul", "preds")?;
    op.add_input(X_test_const);
    op.add_input(theta);
    op.finish()?
  };

  // remaining code ...
```

Now since the whole graph is built, we can create a graph session that will do the actual computation (Listing 2-19).

Listing 2-19. chapter2/rust_and_tf/src/linear_regression.rs

```rust
pub fn run() -> Result<(), Box<dyn Error>> {
  // previous code ...

  let session = Session::new(&SessionOptions::new(),
    &graph)?;
  let mut args = SessionRunArgs::new();
  let preds_token = args
    .request_fetch(&predictions, 0);
  session.run(&mut args)?;
  let preds_token_res: Tensor<f64> = args.fetch::<f64>(
    preds_token)?;
  println!("r-squared error score: {:?}",
    r_squared_score(&preds_token_res.to_vec(),
    &boston_y_test));

  Ok(())
}
```

The r_squared_score is taken from the ml-utils library, and we will discuss the function in the model evaluation section.

The output for the code in Listing 2-19 should be similar to Listing 2-20.

Listing 2-20. chapter2/rust_and_tf output

```
$ cd chapter2/rust_and_tf
$ cargo run lr
r-squared error score: 0.6779404103641853
```

Notice that in this code we create a session, pass the relevant arguments, and then get the values from the output tensor.

We can run this module using cargo run lr and that should give the output.

A different approach that is recommended by the Rust Tensorflow team is to create the computational graph using python and saving the model in a pb file. Once done, you can load the model in a Rust program and then run the computations. Once a computational graph is created, we can save it using the code similar to Listing 2-21.

Listing 2-21. chapter2/rust_and_tf/tensorflow%20create%20model. ipynb

```python
directory = 'boston_regression'
builder = SavedModelBuilder(directory)

with tf.Session(graph=tf.get_default_graph()) as sess:
  sess.run(init)

  signature_inputs = {
    "x": build_tensor_info(X),
    "y": build_tensor_info(Y)
  }
  signature_outputs = {
    "out": build_tensor_info(y_preds)
  }
  signature_def = build_signature_def(
    signature_inputs, signature_outputs,
    REGRESS_METHOD_NAME)
  builder.add_meta_graph_and_variables(
    sess, [TRAINING, SERVING],
    signature_def_map={
      REGRESS_METHOD_NAME: signature_def
    },
    assets_collection=tf.get_collection(
      tf.GraphKeys.ASSET_FILEPATHS))
builder.save(as_text=False)
```

Now once the model is saved, we should be able to load and run the model in our Rust program. Note that the function would have similar data loading and partitioning to test train similar to what we have seen in Listings 2-9 and previous ones.

```rust
pub fn run() -> Result<(), Box<dyn Error>> {
  // previous code ...

  let export_dir = "boston_regression/"; // y = w * x + b

  let mut graph = Graph::new();
  let session = Session::from_saved_model
    (&SessionOptions::new(), &["train", "serve"],
    &mut graph, export_dir)?;

  let op_x = graph
    .operation_by_name_required("x")?;
  let op_x_test = graph
    .operation_by_name_required("x_test")?;
  let op_y = graph
    .operation_by_name_required("y")?;
  let op_train = graph
    .operation_by_name_required("train")?;
  let op_w = graph
    .operation_by_name_required("w")?;
  let op_y_preds = graph
    .operation_by_name_required("y_preds")?;

  Session::new(&SessionOptions::new(), &graph)?;
  let mut args = SessionRunArgs::new();
  args.add_feed(&op_x, 0, &X_train);
  args.add_feed(&op_x_test, 0, &X_test);
  args.add_feed(&op_y, 0, &y_train);
  args.add_target(&op_train);
```

```
let preds_token = args.request_fetch(
  &op_y_preds, 0);
for _ in 0..10 {
    session.run(&mut args)?;
};
let preds_token_res: Tensor<f64> = args
  .fetch::<f64>(preds_token)?;
println!("{:?}", &preds_token_res[..]);
println!("{:?}", &boston_y_test);
println!("{:?}", r_squared_score(
  &preds_token_res[..], &boston_y_test));

// remaining code ...
```

Notice in this code the appropriate variables are called using the Tensorflow names.

This should save your model in the folder `boston_regression/saved_model.pb`. We should now be able to run the code and get the results.

2.5 Gaussian Process

Gaussian Processes can be used in regression problems. One can think of a Gaussian process as defining distribution over functions, and inference taking place directly in the space of functions, the function-space view. The predictions from a Gaussian process model take the form of a full predictive distribution. Thus, a Gaussian process is a collection of random variables, any finite number of which have a joint Gaussian distribution.

A Gaussian process is completely specified by its mean function and covariance function. Thus, in a Gaussian process, the priors need to be specified. Generally, the prior mean is assumed to be zero and the covariance is specified by passing a kernel object.

The noise levels of the Gaussian process can also be passed. A moderate noise level can be helpful for dealing with numeric issues when fitting as it is effectively implemented as Tikhonov regularization, that is, by adding it to the diagonal of the kernel matrix.

Create a Gaussian process model to train it on the training part of the boston dataset above. Here a kernel with lengthscale 2 and amplitude 1 is defined as the covariance function. A Zero function is defined as the mean function and noise levels is clipped at 10.0. We then pass the training xâ™s and yâ™s to the train method. This optimizes the model according to the training values (Listing 2-22).

Listing 2-22. chapter2/rustymachine_regression/src/gaussian_
process_reg.rs

```
use rusty_machine::learning::gp::GaussianProcess;
use rusty_machine::learning::gp::ConstMean;
use rusty_machine::learning::toolkit::kernel;
use rusty_machine::learning::SupModel;

pub fn run() -> Result<(), Box<dyn Error>> {
  // previous data loading and partitioning code ...

  // the kernel function
  let ker = kernel::SquaredExp::new(2., 1.);
  // the mean function
  let zero_mean = ConstMean::default();

  // model definition
  let mut gaus_model = GaussianProcess::new(ker, zero_mean,
  10f64);
  gaus_model.train(&boston_x_train,
    &boston_y_train)?;

  let predictions = gaus_model.predict(&boston_x_test).unwrap();
  let predictions = Matrix::new(test_size, 1, predictions);
  let acc = neg_mean_squared_error(&predictions, &boston_y_test);
  println!("gaussian process regression error: {:?}", acc);
```

```
println!("gaussian process regression R2 score: {:?}",
r_squared_score(
  &boston_y_test.data(), &predictions.data()));

Ok(())
}
```

Running this code without any errors would mean that we are able to create the model successfully. We should get an output similar to Listing 2-23.

Listing 2-23. chapter2/rustymachine_regression/src/gaussian_ process_reg.rs output

```
$ cd chapter2/rustymachine_regression
$ cargo run gp
gaussian process regression error: -593.9085507602914
gaussian process regression R2 score: -5.578006698530541
```

2.6 Generalized Linear Models

Ordinary linear regression assumed that the unknown quantity is the linear combination of the set of predictors. This assumption is fine if the response variable is a normal distribution of the input variables. Such data can be any kind of data that is relatively stable and varies by a small amount. This can be seen in many natural distributions such as the heights of human beings where one value is independent of another value and hence the Central Limit Theorem is able to play a role. In real life though, many distributions are not normal.

The generalized linear model is a flexible variation of the OLS Regression model that allows for response variables that have error distribution models other than normal distribution. In these models, the response variable y_i is assumed to follow an exponential family distribution of mean μ_i, which is assumed to be some (linear or nonlinear) function of $x_i^T \beta$ [1].

The GLM consists of three elements [2].

1. A linear predictor similar to OLS

$$\eta_i = \beta_0 + \beta_1 x_{1i} + \ldots + \beta_p x_{pi} \qquad (2)$$

   or in matrix form $\eta = X\beta$

2. A link function that describes how the mean $E(Y_i) = \mu_i$ depends on the linear predictor

$$g(\mu_i) = \eta_i \qquad (3)$$

3. A variance function that describes how the variance, var(Y_i) depends on the mean

$$\mathrm{var}(Y_i) = \phi V(\mu) \qquad (4)$$

It is convenient if V follows an exponential family of distributions. The unknown parameters β are typically estimated using maximum likelihood, maximum quasi-likelihood, or Bayesian methods. It can be shown that the general linear model is a special case of the GLM where the link function is $g(\mu_i) = \mu_i$ and the variance function is $V(\mu_i) = 1$.

In rusty_machine we can create a linear regression model using the code shown in Listing 2-24.

Listing 2-24. chapter2/rustymachine_regression/src/glms.rs

```
use rusty_machine::learning::glm::{GenLinearModel, Normal};

pub fn run() -> Result<(), Box<dyn Error>> {
  // starting code for data loading and partitioning...
```

```
// Create a normal generalised linear model
let mut normal_model = GenLinearModel::new(Normal);
normal_model.train(
  &boston_x_train, &boston_y_train)?;

let predictions = normal_model.predict(
  &boston_x_test).unwrap();
let predictions = Matrix::new(
  test_size, 1, predictions);
let acc = neg_mean_squared_error(
  &predictions, &boston_y_test);
println!("glm poisson accuracy: {:?}", acc);
  println!("glm poisson R2 score: {:?}", r_squared_score(
      &boston_y_test.data(), &predictions.data()));
Ok(())
}
```

Output for this code should be similar to Listing 2-25.

Listing 2-25. glms.rs output

```
$ cargo run glms
glm poisson accuracy: -36.55900309522449
glm poisson R2 score: 0.6478692745681216
```

Other models that are implemented are Bernoulli and Binomial, which are mainly used for classification. We can also use Poisson regression for the Criterion. Poisson regression is similar to multiple regression except that the dependent variable is an observed count that follows the Poisson distribution. Thus the possible values of Y are non-negative integers: 0,1,... and so on. It is assumed that the large counts are rare.

2.7 Evaluation of Regression Models

So far in all the code for the machine learning models that we have discussed so far, we have used a couple of functions such as neg_mean_squared_error or r_squared_score, which we have avoided talking about so far. These functions were used to evaluate how good the models that were created were, and in this section, we will discuss those functions, their creation, and usage in detail.

Evaluating a model is important to understand if it is doing a good job of predicting the target on new and future data. Because future instances have unknown target values, we need to check the accuracy metric of the ML model on the data for which we already know. That is the reason that before training, we split the dataset that we have into train and test sets. This assessment acts as a proxy for evaluating how close the ML model is to mimicking the real-world use case.

2.7.1 MAE and MSE

For continuous variables, two of the most popular methods of evaluation metrics are Mean Absolute Error and Mean Square Error [3]. Mean Absolute Error is the average of the absolute difference between the predicted values and observed values. For a vector of n predictions $\hat{y}$ generated from a sample of n data points and if y is the observed values for those data points, then

$$\in_{MAE} = \frac{1}{n} \sum_{i=1}^{n} \left| y_i - \hat{y}_i \right| \tag{5}$$

Mean Squared Error is the summation of the square of the differences between the predicted and observed values (also called residuals).

$$\in_{MSE} = \frac{1}{n} \sum_{i=1}^{n} \left(y_i - \hat{y}_i \right)^2 \tag{6}$$

In many cases the root of the above is also taken, $\sqrt{\in_{MSE}}$. This essentially transforms it to a sample standard deviation and is called the root-mean-squared-error.

Now RMSE is the best error estimator when it comes to Gaussian distributions. Sadly, most of the real-world use cases are not strictly Gaussian. They act Gaussian-like only in a limited space. The result is that we give undue importance to the outliers due to taking the square of the residuals. MAE is also easier to understand and does not give too much importance to outliers. However, RMSE is the default metric of many models because loss function defined in terms of RMSE is smoothly differentiable and makes it easier to perform mathematical operations needed in machine learning.

In rusty machine to get the mean square error of these models, once the training is done, we pass boston_x_test as a reference to predict the method of the model to get the predictions. These predictions are compared with the actual values to get the neg_mean_squared_error function. This function returns the additive inverse of the mean-squared error. Mean square error is the average of the squared differences between prediction and actual observation (Listing 2-26).

Listing 2-26. chapter2/rustymachine_regression/src/lin_reg.rs

```
use rusty_machine::analysis::score::neg_mean_squared_error;

pub fn run() -> Result<(), Box<dyn Error>> {
    // previous part ...

    let predictions = lin_model.predict(&boston_x_test).unwrap();
    let predictions = Matrix::new(test_size, 1, predictions);
    let acc = neg_mean_squared_error(&predictions, &boston_y_test);
    println!("linear regression error: {:?}", acc);
```

```
let predictions = gaus_model.predict(&boston_x_test).unwrap();
let predictions = Matrix::new(test_size, 1, predictions);
let acc = neg_mean_squared_error(&predictions, &boston_y_test);
println!("gaussian process regression error: {:?}", acc);

// remaining code...
```

2.7.2 R-Squared Error

R-squared evaluates the scatter of the data points around the fitted regression line. It is also called the coefficient of determination. For the same dataset, higher R-squared values represent smaller differences between the observed and fitted values.

R-squared is the percentage of the dependent variable variation that the linear model explains.

$$R^2 = \frac{\text{Variance explained by the model}}{Total\ variance} \tag{7}$$

$$= 1 - \frac{\frac{1}{n}\sum_{i=1}^{n}(y_i - \hat{y}_i)^2}{\frac{1}{n}\sum_{i=1}^{n}(y_i - \bar{y})^2} \tag{8}$$

We can see in Figure 2-2 a visual demonstration of how R-squared values represent the scatter around the regression line, using two sample data points.

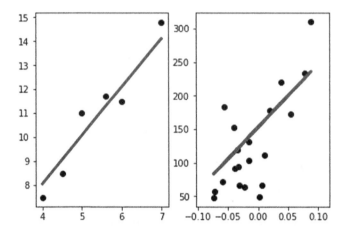

Figure 2-2. *Sample regression fits*

The r-squared for the regression model on the left is 93% and for the model on the right is 47%. When a regression model accounts for more of the variance, the data points are closer to the regression line. In practice, a regression model with R^2 of 100% is never observed. In that case, fitted values equal the data values and consequently all observations fall exactly on the regression line.

The next function can be used as a reference to implement r-squared in Rust. We first take the model variance, which is the sum of the squares of the consecutive difference between actual and predicted values. Then we take the mean of the actual distribution and use that to calculate the variance of the test distribution. Then we run it through the r-squared formula (Listing 2-27).

Listing 2-27. ml-utils/src/sup_metrics.rs

```
fn r_squared_score(y_test: &Vec<f64>, y_preds: &Vec<f64>) -> f64 {
  let mv: f64 = y_test.iter().zip(y_preds.iter()).fold(
    0., |v, (y_i, y_i_hat)| {
      v + (y_i - y_i_hat).powi(2)
    }
  );
```

```
let mean = y_test.iter().sum::<f64>() as f64
  / y_test.len() as f64;

let var =  y_test.iter().fold(
  0., |v, &x| {v + (x - mean).powi(2)}
);
let r2: f64 = 1.0 - (mv / var);
r2
}
```

This function should not be usable by passing the y_test and y_pred to get the score. Since, in the examples in this chapter, the y_test and predicted values are in rusty machine Matrices, we will probably need to do something like that shown in Listing 2-28.

Listing 2-28. chapter2/rustymachine_regression/src/glms.rs

```
pub fn run() -> Result<(), Box<dyn Error>> {
  // previous part of the function...

  println!("glm poisson R2 score: {:?}", r_squared_score(
    &boston_y_test.data(), &predictions.data()));
}
```

2.8 Classification Algorithms

Scenarios where the machine learning algorithm is tasked with bucketing input variables to predefined buckets are called classification problems. In this section, we will be creating classification models in Rust.

In Rust we can create classifiers using the package rustlearn. This crate contains effective implementation for a number of common machine learning algorithms. To be able to use rustlearn, we will need to implement the floats as float32.

2.8.1 Iris Dataset

For showing the usage of the classification algorithms, we will be using the Iris dataset.[4] Download the dataset from the website or find it in the code that is shared with this book in the datasets folder. It is a multivariate dataset with the following features.

- sepal length in cm.

- sepal width in cm.

- petal length in cm.

- petal width in cm.

- classes: setosa, versicolor, and virginica.

The code that is explained here is kept in the rustlearn_ classification_tasks folder. Inside the package, create a folder data and keep the csv file there (Listing 2-29).

Listing 2-29. Iris data

```
$ head -n2 data/iris.csv
sepal_length,sepal_width,petal_length,petal_width,species
5.1,3.5,1.4,0.2,setosa
$ wc -l data/iris.csv
151 data/iris.csv # there are 150 samples in the dataset.
```

Similar to before, we are going to work with rustlearn, csv, rand, and serde. Rustlearn crate is a new dependency that you are seeing now. We will be using this crate for the classification tasks and will be talking about it soon. Find these dependencies in the Cargo.toml file (Listing 2-30).

---

[4]https://archive.ics.uci.edu/ml/datasets/iris.

Listing 2-30. chapter2/rustlearn_classification_tasks/Cargo.toml

```
[package]
name = "rustlearn_classification_tasks"
version = "0.1.0"
edition = "2018"

[dependencies]
rustlearn = "0.5.0"
csv = "1.0.5"
serde = "1.0.89"
serde_derive = "1.0.89"
rand = "0.6"
ml-utils = { path = "../ml-utils" }
```

Remember that in the regression case, we created a `BostonHousing` struct. In this case we are going ahead with creating a `Flower struct`. Notice that the code structures will be largely similar (Listing 2-31).

Listing 2-31. chapter2/ml-utils/src/datasets.rs

```
extern crate serde;
#[macro_use]
extern crate serde_derive;

#[derive(Debug, Deserialize)]
pub struct Flower {
  sepal_length: f32, sepal_width: f32,
  petal_length: f32, petal_width: f32,
  species: String,
}
```

We are going with the f32 datatype now as we will be using the rustlearn package, which requires the inputs to be in f32 floats.

Here species is the label and other columns are the features. So, to parse out the features and the labels we will implement into_feature_ vector and into_labels for the Flower struct. Note that in case of defining the feature vector, the order of the placement in the csv file is maintained. And in case of label encoding some numbers (which are arbitrary) are given to the different labels. Ideally along with a label encoding method, a label decoder should also be implemented for the final response. In this case for the sake of brevity that is not shown (Listing 2-32).

Listing 2-32. ml-utils/src/datasets.rs

```
use std::io; use std::vec::Vec; use csv;
impl Flower {
  pub fn into_feature_vector(&self) -> Vec<f32> {
    vec![self.sepal_length, self.sepal_width,
    self.petal_length, self.petal_width]
  }

  pub fn into_labels(&self) -> f32 {
    match self.species.as_str() {
      "setosa" => 0., "versicolor" => 1.,
      "virginica" => 2.,
      some_other => panic!("Not able to parse the label.
        Some other label got passed. {:?}", some_other),
    }
  }
}
```

We will use the amazing stdin and read the package using a command similar to cargo run lr < data/iris.csv. Here cargo run will compile and run the binary and lr is the argument after that which is enabled in the package. Take a look at main.rs for all the options. So to enable that we will use the `std::io` package and pass it to a `rdr` variable using `csv::Reader`. We now serialize each record to the `Flower` struct and push it to a data vector, a similar data vector that we have seen in the regression section. With this we will also shuffle the data for good measure (Listing 2-33).

Listing 2-33. chapter3/rustlearn_classification_tasks/src/logistic_reg.rs

```
use ml_utils::datasets::Flower;

use csv; use rand::thread_rng;
use rand::seq::SliceRandom;

pub fn run() -> Result<(), Box<dyn Error>> {
  let mut rdr = csv::Reader::from_reader(io::stdin());
  let mut data = Vec::new();
  for result in rdr.deserialize() {
    let r: Flower = result?; // should have Box<dyn Error> in
                                     the defn.
    data.push(r);
  }
  data.shuffle(&mut thread_rng());

  // rest of the code...
```

Now we will be separating out the train and test datasets. The code is similar to the one in the regression section except here in this case we will need to convert the vectors into `rustlearn` sparse or dense vectors (Listing 2-34).

Listing 2-34. chapter3/rustlearn_classification_tasks/src/logistic_
reg.rs

```
use rustlearn::prelude::*;

pub fn run() -> Result<(), Box<dyn Error>> {
  // previous part of the function ...

  // separate out to train and test datasets.
  let test_size: f32 = 0.2;
  let test_size: f32 = data.len() as f32 * test_size;
  let test_size = test_size.round() as usize;
  let (test_data, train_data) = data.split_at(test_size);
  let train_size = train_data.len();
  let test_size = test_data.len();

  // differentiate the features and the labels.
  let flower_x_train: Vec<f32> = train_data.iter()
    .flat_map(|r| r.into_feature_vector()).collect();
  let flower_y_train: Vec<f32> = train_data.iter()
    .map(|r| r.into_labels()).collect();
  let flower_x_test: Vec<f32> = test_data.iter()
    .flat_map(|r| r.into_feature_vector()).collect();
  let flower_y_test: Vec<f32> = test_data.iter()
    .map(|r| r.into_labels()).collect();

  // Convert the vectors to a dense matrix or a sparse matrix
  let mut flower_x_train = Array::from(flower_x_train);
  // reshape so that the read training vector is currently a
    flat error.
  flower_x_train.reshape(train_size, 4);
  let flower_y_train = Array::from(flower_y_train);
  let mut flower_x_test = Array::from(flower_x_test);
```

```
// Similarly the test vector also needs to be reshaped.
flower_x_test.reshape(test_size, 4);

// rest of the function ...
```

We should now be able to train the data on `rustlearn` models.

2.8.2 Logistic Regression

Logistic Regression is a popular classification technique in which a logit function is used to model a binary dependent variable. The assumption for the dependent variable is that it follows Bernoulli distribution. While OLS regression is a distance-minimizing approximation method, estimation of parameters is done using the maximum likelihood method. Maximizing the likelihood function determines the parameters that are most likely to produce the observed data.

Unlike regression, for normally distributed residuals, it is not possible to find a closed form solution that maximizes the function. So an iterative approach has to be used. One of the popular iterative approaches is Stochastic Gradient Descent (SGD). SGD is implemented in `rustlearn` and to implement model training in `rustlearn`, we call the `Hyperparameter` module from `linear_models` and can pass various parameters such as learning rate, l1 and l2 penalties, and if it is a multi-label classification or binary classification (Listing 2-35).

Listing 2-35. chapter3/rustlearn_classification_tasks/src/logistic_reg.rs

```
pub fn run() -> Result<(), Box<dyn Error>> {
  // previous part of the fn ..

  let mut model = lr::new(4)
    .learning_rate(0.1).l2_penalty(0.5)
    .l1_penalty(0.0).one_vs_rest();
```

```
for _ in 0..100 { // for 100 epochs
  model.fit(&flower_x_train, &flower_y_train).unwrap();

let prediction = model.predict(&flower_x_test).unwrap();
let acc1 = accuracy_score(&flower_y_test, &prediction);
println!("Logistic Regression: accuracy: {:?}", acc1);

Ok(())
}
```

Running the for loop on the model is equivalent to training the model for multiple epochs.

Running the mode in Listing 2-35 should give an output similar to that shown in Listing 2-36.

Listing 2-36. logistic_reg.rs output

```
$ cargo run lr < ../datasets/iris.csv
Logistic Regression: accuracy: 0.36666667
```

2.8.3 Decision Trees

Let us try to redefine the classification problem and understand it from a different perspective. In a classification problem, we have a training sample of n observations on a class variable Y that takes values 1, 2, ..., k and p predictor variables, $X_1, X_2, ..., X_p$. where x_m is the training data in the node m. Our goal is to find a model for predicting the values of Y for new X values. We can think of this problem as simply a partition of the X-space into k disjoint sets, $A_1, A_2, ..., A_k$, such that the predicted value of Y is j if X belongs to A_j, for $j = 1, 2, ..., k$. Classification trees take this approach. They yield rectangular sets A_j by recursively partitioning the dataset one X variable at a time. This makes the sets easier to interpret. A key advantage of the tree structure is its applicability to any number of variables.

The key idea is this:

1. Grow an overly large tree by using forward selection. At each step, find the best split. Grow until all terminal nodes either

    (a)   have $<n$ data points,

    (b)   all nodes in a node have the same outcome. If this happens the node is said to be "pure."

2. Prune the tree back, creating a nested sequence of trees, decreasing in complexity.

Implementing decision trees in rustlearn can be done with the code shown in Listing 2-37. It supports the CART algorithm for both dense and sparse data and features are selected using reduction in Gini impurity [4].

Listing 2-37. chapter3/rustlearn_classification_tasks/src/trees.rs

```
use rustlearn::trees::decision_tree;

pub fn run() -> Result<(), Box<dyn Error>> {
  // data loading and transformations part ...
  // similar to the logistic regression secion above ...

  let mut decision_tree_model = decision_
  tree::Hyperparameters::new(
    flower_x_train.cols()).one_vs_rest();
  decision_tree_model.fit(&flower_x_train, &flower_y_train).
  unwrap();

  let prediction = decision_tree_model.predict(
    &flower_x_test).unwrap();
```

```
  let acc = accuracy_score(
    &flower_y_test, &prediction);
  println!("DecisionTree model accuracy: {:?}", acc);

  Ok(())
}
```

Running the above code should give an output similar to that shown in Listing 2-38.

Listing 2-38. chapter3/rustlearn_classification_tasks/src/trees.rs output

```
$ cargo run trees < ../datasets/iris.csv
    Finished dev [unoptimized + debuginfo] target(s) in 0.03s
     Running `target/debug/rustlearn_classification_tasks trees`
DecisionTree model accuracy: 0.96666664
```

2.8.4 Random Forest

An improvement of decision trees is the Random Forest. In this case, an ensemble of trees is grown, and voting is done among them to get the most popular class. The trees are a combination of tree predictors such that each tree depends on the values of a random vector sampled independently and with the same distribution for all trees in the forest. The generalization for the forests converges to a limit as the number of trees in the forest becomes large [4].

To implement random forests in rustlearn, we will need to pass the previous decision trees variable with the parameters to the Random Forest hyperparameter module so that a collection of trees can be built (Listing 2-39).

Listing 2-39. chapter3/rustlearn_classification_tasks/src/trees.rs

```rust
use rustlearn::ensemble::random_forest::Hyperparameters as rf;

pub fn run() -> Result<(), Box<dyn Error>> {
  // previous part of the fn ...

  let mut tree_params = decision_tree::Hyperparameters::new(
    flower_x_train.cols());
  tree_params.min_samples_split(10)
    .max_features(4);
  let mut model = randomforest::new(
    tree_params, 10).one_vs_rest();
  model.fit(&flower_x_train,
            &flower_y_train).unwrap();

  let prediction = random_forest_model
    .predict(&flower_x_test).unwrap();
  let acc = accuracy_score(
    &flower_y_test, &prediction);
  println!("Random Forest: accuracy: {:?}", acc);

  Ok(())
}
```

In the example, the above code is below the code for the decision trees and hence in the output we will see the results for both the codes (Listing 2-40).

Listing 2-40. chapter3/rustlearn_classification_tasks/src/trees.rs output

```
$ cargo run trees < ../datasets/iris.csv
    Finished dev [unoptimized + debuginfo] target(s) in 0.03s
     Running `target/debug/rustlearn_classification_tasks trees`
DecisionTree model accuracy: 1.0
Random Forest: accuracy: 1.0
```

> **Note**   Notice that when we had started with logistic regressions and other simpler algorithms, the accuracy was quite low, around 30%; but as we have progressed to tree-based models, we have reached accuracy of around 95–100%. Keep in mind that these are toy datasets and have been chosen because they are well known, simple, and because the focus of this book is to show Rust capabilities in machine learning. Be wary of such accuracy levels in real-world problems.

2.8.5 XGBoost

To get even better at creating a good classification model using trees, one idea is to assemble a series of weak learners and convert them into a strong classifier. XGBoost means Extreme Gradient Boosting, so let's take those terms apart one by one. In boosting, the trees are built sequentially so that each subsequent tree aims to reduce the errors of the previous tree. Each tree learns from its predecessors and updates the residual errors. Hence the tree that learns next in the sequence will learn from an updated version of the residuals. This results in really strong classifiers [5].

In contrast to Random Forest, where the trees are grown to the maximum extent, boosting makes use of trees with fewer splits. Such small trees, which are not very deep, are highly interpretable. Cross-validation is an important step in boosting as this is used to find optimal parameters such as the number of tress or iterations, the rate at which the gradient boosting learns, and the depth of the tree.

Parameters Before running XGBoost, we must set three types of parameters: general parameters, booster parameters, and task parameters [6].

In this case, instead of `rustlearn` we will be using the `rust-xgboost` library.[5] The full code is kept in the package `iris_classification_xgboost`. for reference. This library is a wrapper around the XGBoost library.[6]

In this case we will need to have `xgboost = "0.1.4,"` which is a wrapper over `XGBoost-v0.82`. In the `main.rs` we will call the relevant modules (Listing 2-41).

Listing 2-41. chapter3/iris_classification_xgboost/src/main.rs

```
use xgboost;
use xgboost::{parameters, DMatrix, Booster};
```

The other data preprocessing steps are the same as we have used in the previous sections for logistic regression for the Iris dataset. Once the vectors `flower_x_train`, `flower_y_train`, `flower_x_test`, and `flower_y_test` vectors are created, we will need to convert it to XGBoost-compatible vectors though. The x vectors are converted to `DMatrix` and the corresponding labels are set (Listing 2-42).

Listing 2-42. chapter3/iris_classification_xgboost/src/main.rs

```
fn read_csv() -> Result<(), Box<dyn Error>> {
  // previous data loading and splitting code ...

  let mut dtrain = DMatrix::from_dense(&flower_x_train, train_
  size).unwrap();

  dtrain.set_labels(&flower_y_train).unwrap();

  let mut dtest = DMatrix::from_dense(&flower_x_test, test_
  size).unwrap();
  dtest.set_labels(&flower_y_test).unwrap();

  // remaining part of the code ...
```

---

[5]Rust XGBoost Source.
[6]Original C++ based XGBoost source.

Now we will set the XGBoost parameters. First the objective function will be set to Multilabel Softmax as the number of labels are more than two. Then we will set the tree-based learning model's parameter. These are utilized in setting the booster configuration (Listing 2-43).

Listing 2-43. chapter3/iris_classification_xgboost/src/main.rs

```
fn read_csv() -> Result<(), Box<dyn Error>> {
  // previous part of the fn ..

  let lps = parameters::learning::LearningTaskParametersBuilder
  ::default()
    .objective(parameters::learning::Objective::MultiSoftmax(3))
    .build().unwrap();
  let tps = parameters::tree::TreeBoosterParametersBuilder::
  default()
    .max_depth(2).eta(1.0)
    .build().unwrap();
  let bst_parms = parameters::BoosterParametersBuilder::default()
    .booster_type(parameters::BoosterType::Tree(tps))
    .learning_params(learning_params)
    .verbose(true).build().unwrap();

  // remaining part of the fn ...
```

After that we will specify which matrices are for training and which for testing. This will be passed to a configuration object that will be utilized during the training (Listing 2-44).

Listing 2-44. chapter3/iris_classification_xgboost/src/main.rs

```rust
fn read_csv() -> Result<(), Box<dyn Error>> {
  // previous code ...

  let ev = &[(&dtrain, "train"), (&dtest, "test")];
  let params = parameters::TrainingParametersBuilder::default()
    .dtrain(&dtrain).boost_rounds(2)
    .booster_params(bst_parms)
    .evaluation_sets(Some(ev))
    .build().unwrap();
  let booster = Booster::train(&params).unwrap();

  // remaining code ...
```

Now the predictions can be made and compared with the actual values (Listing 2-45).

Listing 2-45. chapter3/iris_classification_xgboost/src/main.rs

```rust
fn read_csv() -> Result<(), Box<dyn Error>> {
  // previous code ...

  let preds = booster.predict(&dtest).unwrap();
  let labels = dtest.get_labels().unwrap();

  // find the accuracy
  let mut hits = 0;
  let mut correct_hits = 0;
  for (predicted, actual) in preds.iter().zip(labels.iter()) {
    if predicted == actual {
      correct_hits += 1;
    }
    hits += 1;
  }
```

```
  assert_eq!(hits, preds.len());
  println!("accuracy={} ({}/{} correct)",
    correct_hits as f32 / hits as f32, correct_hits, preds.len());
}
```

The output for the above code should be similar to that below with accuracy of around 93–96% (Listing 2-46).

Listing 2-46. chapter3/iris_classification_xgboost/src/main.rs

```
$ cd chapter2/iris_classification_xgboost
$ cargo run < ../datasets/iris.csv
    Finished dev [unoptimized + debuginfo] target(s) in 0.05s
    Running `target/debug/iris_classification_xgboost`
[08:26:11] DANGER AHEAD: You have manually specified `updater`
parameter. The `tree_method` parameter will be ignored.
Incorrect sequence of updaters will produce undefined behavior.
For common uses, we recommend using `tree_method` parameter
instead.
[08:26:11] src/tree/updater_prune.cc:74: tree pruning end, 1
roots, 2 extra nodes, 0 pruned nodes, max_depth=1
[08:26:11] src/tree/updater_prune.cc:74: tree pruning end, 1
roots, 4 extra nodes, 0 pruned nodes, max_depth=2
[08:26:11] src/tree/updater_prune.cc:74: tree pruning end, 1
roots, 6 extra nodes, 0 pruned nodes, max_depth=2
[0]     test-merror:0  train-merror:0.033333
[08:26:11] src/tree/updater_prune.cc:74: tree pruning end, 1
roots, 2 extra nodes, 0 pruned nodes, max_depth=1
[08:26:11] src/tree/updater_prune.cc:74: tree pruning end, 1
roots, 6 extra nodes, 0 pruned nodes, max_depth=2
[08:26:11] src/tree/updater_prune.cc:74: tree pruning end, 1
roots, 6 extra nodes, 0 pruned nodes, max_depth=2
[1]     test-merror:0  train-merror:0.011111
```

```
preds: [1.0, 2.0, 0.0, 2.0, 1.0, 0.0, 0.0, 0.0, 0.0, 1.0, 2.0,
0.0, 2.0, 0.0, 1.0, 0.0, 2.0, 2.0, 0.0, 1.0, 2.0, 0.0, 1.0,
1.0, 1.0, 2.0, 0.0, 2.0, 0.0, 0.0]
[1.0, 2.0, 0.0, 2.0, 1.0, 0.0, 0.0, 0.0, 0.0, 1.0, 2.0, 0.0,
2.0, 0.0, 2.0, 0.0, 2.0, 2.0, 0.0, 2.0, 2.0, 0.0, 1.0, 1.0,
1.0, 2.0, 0.0, 2.0, 0.0, 0.0]
accuracy=0.93333334 (28/30 correct)
```

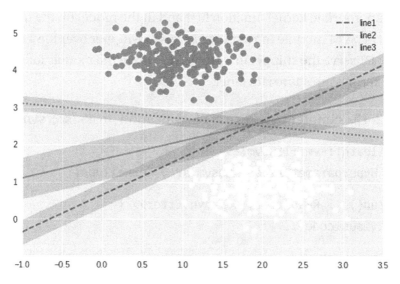

Figure 2-3. *Svm*

2.8.6 Support Vector Machines

A Support Vector Machine (SVM) is a discriminative classifier formally
defined by a separating hyperplane. Possible hyperplanes are shown
in Figure 2-3. In other words, given labeled training data, the algorithm
outputs an optimal hyperplane that categorizes new examples. In two-
dimensional space, this hyperplane is a line dividing a plane into two
parts wherein each class lays on either side. The vectors that define the
hyperplane are the support vectors.

77

Kernel functions change the definition of the dot product in the linear formulation. The different types of common dot products are linear, polynomial, radial basis functions, and sigmoid functions. Generally RBF functions are taken as the default kernels in most svm models [7].

In Figure 2-1 with different data points, multiple hyperplanes are experimented with but the hyperplane (line2) is chosen because it has the highest degree of margin.

To implement SVM in `rustlearn`, we will need to create the model with the appropriate kernel function first and fit the model on the training dataset. The SVM module in `rustlearn` is just a wrapper over the `libsvm` package.[7] Observe the snippet in Listing 2-47 where the models for the different kernels are shown together.

Listing 2-47. chapter3/rustlearn_classification_tasks/src/svm.rs

```
use rustlearn::svm::libsvm::svc::{
        Hyperparameters as libsvm_svc, KernelType};

pub fn run() -> Result<(), Box<dyn Error>> {
  // previous code ...

  let svm_linear_model = libsvm_svc::new(
    4, KernelType::Linear, 3)
    .C(0.3).build();
  let svm_poly_model = libsvm_svc::new(4, KernelType::
Polynomial, 3)
    .C(0.3) .build();
  let svm_rbf_model = libsvm_svc::new(4, KernelType::RBF, 3)
    .C(0.3).build();
  let svm_sigmoid_model = libsvm_svc::new(4, KernelType::Sigmoid, 3)
    .C(0.3).build();
```

---

[7]https://www.csie.ntu.edu.tw/~cjlin/libsvm/.

```
let svm_kernel_types = ["linear", "polynomial", "rbf",
"sigmoid"];
let mut svm_model_types = [svm_linear_model, svm_poly_model,
svm_rbf_model, svm_sigmoid_model];
for (kernel_type, svm_model) in svm_kernel_types.iter().
zip(svm_model_types.iter_mut()) {
    svm_model.fit(&flower_x_train, &flower_y_train).unwrap();

    let prediction = svm_model.predict(&flower_x_test).unwrap();
    let acc = accuracy_score(&flower_y_test, &prediction);
    println!("Lib svm {kernel}: accuracy: {accuracy}",
    accuracy=acc, kernel=kernel_type);
};

// remaining code ...
```

We should be getting an output similar to that shown in Listing 2-48.

Listing 2-48. output

```
Lib svm linear: accuracy: 0.9
Lib svm polynomial: accuracy: 0.9
Lib svm rbf: accuracy: 0.8666667
Lib svm sigmoid: accuracy: 0.26666668
```

2.8.7 K Nearest Neighbors

For classification in Rust, we have mostly been focused on `rustlearn`, and for regression we have used `rusty machine`, but one of the popular classifiers that is available in rusty machine and not in `rustlearn` is the K nearest neighbor's algorithm. The assumption in the KNN algorithm is that "birds of the same feather flock together" or in other words, similar things tend to have the same results. So, an object is classified by the plurality vote of the neighbors.

In rusty machine, the KNN classifiers have not been published on the crate, and hence we will need to clone the repo and build from the path or use the github reference (Listing 2-49).

Listing 2-49. chapter2/rusty_machine_classification/Cargo.toml

```
[package]
name = "rusty_machine_classification"
version = "0.1.0"
edition = "2018"

[dependencies]
rusty-machine = { path = "../rusty-machine" }
ml-utils = { path = "../ml-utils" }
rand = "0.6.5"
csv = "1.0.7"
```

Since this is a classification algorithm, we will be using the Flower struct and Iris dataset. The Flower struct has been implemented for usage with rustlearn and hence we will need to do some extra maneuvering (Listing 2-50).

Listing 2-50. chapter2/rusty_machine_classification/src/main.rs

```
use rusty_machine as rm;
use rm::linalg::Matrix;
use rm::linalg::Vector;

use ml_utils;
use ml_utils::datasets::Flower;

fn main() -> Result<(), Box<dyn Error>> {
  // previous data loading and splitting code ...
```

```rust
// differentiate the features and the labels.
let flower_x_train: Vec<f64> = train_data.iter().flat_map(|r| {
  let features = r.into_feature_vector();
  let features: Vec<f64> = features.iter().map(
    |&x| x as f64).collect();
  features
}).collect();
let flower_y_train: Vec<usize> = train_data.iter().map(
  |r| r.into_int_labels() as usize).collect();

let flower_x_test: Vec<f64> = test_data.iter().flat_map(|r| {
  let features = r.into_feature_vector();
  let features: Vec<f64> = features.iter().map(
    |&x| x as f64).collect();
  features
}).collect();
let flower_y_test: Vec<u32> = test_data.iter().map(
  |r| r.into_int_labels() as u32).collect();

// COnvert the data into matrices for rusty machine
let flower_x_train = Matrix::new(train_size, 4, flower_x_train);
let flower_y_train = Vector::new(flower_y_train);
let flower_x_test = Matrix::new(test_size, 4, flower_x_test);

// remaining code ...
```

Remember that rustlearn needed vectors of f32, while in rusty machine we needed to create rulinalg matrices of f64, hence we need to convert these f32 vectors to f64 and then create the matrices in the above code. Also, the into_int_labels of the Flower struct are the same as into_labels but implemented for u64 as we are dealing with specific labels in this case (Listing 2-51).

Listing 2-51. chapter2/rust-lang/ml-utils/src/datasets.rs

```
impl Flower {
  // ... previous code for Flower

  pub fn into_int_labels(&self) -> u64 {
    match self.species.as_str() {
      "setosa" => 0,
      "versicolor" => 1,
      "virginica" => 2,
      l => panic!(
        "Not able to parse the target.
        Some other target got passed. {:?}", l),
    }
  }
}
```

This was converted to usize as the KNN struct that we are going to use later implemented labels as usize.

We should now be able to create a simple KNN classifier and train it on the Iris dataset (Listing 2-52).

Listing 2-52. chapter2/rusty_machine_classification/src/main.rs

```
use rm::learning::knn::KNNClassifier;
use rusty_machine::learning::knn::{KDTree, BallTree,
BruteForce};
use rm::learning::SupModel;
use ml_utils::sup_metrics::accuracy;

fn main() -> Result<(), Box<dyn Error>> {
  // previous code ...
```

```
let mut knn = KNNClassifier::new(2); // model initialisation
knn.train(&flower_x_train, &flower_y_train).unwrap();
// training

let preds = knn.predict(&flower_x_test).unwrap(); // prediction

let preds: Vec<u32> = preds.data().iter().map(|&x| x as u32).
collect();
println!("accuracy {:?}",
  accuracy(preds.as_slice(), &flower_y_test)); // accuracy

Ok(())
}
```

The accuracy reported should be around 93-100%.

In rusty-machine we can have different KNN models apart from the default one, which is the KDTree algorithm. We can use the Ball tree algorithm, which is used when the number of dimensions are huge, or the Brute force algorithm, the advantage of which is that it is embarrassingly parallelizable (Listing 2-53).

Listing 2-53. chapter2/rusty_machine_classification/src/main.rs

```
use rusty_machine::learning::knn::{KDTree, BallTree, BruteForce};

fn main() -> Result<(), Box<dyn Error>> {
  // previous code ...

  let mut knn = KNNClassifier::new_specified(2, BallTree::new(30));

  let mut knn = KNNClassifier::new_specified(2, KDTree::default());

  let mut knn = KNNClassifier::new_specified(2, BruteForce::
default());

  // remaining code ...
```

For the whole code, take a look at the `rusty_machine_classification` package in the chapter2 folder.

2.8.8 Neural Networks

Neural networks are a set of algorithms, modeled loosely after the human brain, which are designed to recognize patterns. One of the most popular ways of using neural networks is by grouping them in stacks. Usable networks that work are seen to be composed of several layers. The layers are made of nodes. A node is just a place where computation happens. A node combines input from the data with a set of coefficients or weights, which either amplify or dampen that input, thereby assigning significance to inputs with regard to the task the algorithm is trying to learn: for example, which input is most helpful in classifying the data without error. These input weight products are summed and passed through an "activation function." Activation functions are generally nonlinear and they work to determine whether and to what extent that signal should progress further through the network to affect the ultimate outcome, an example being the classification task. If the signal passes through, the neuron has been "activated." Take a look at Figure 2-4 to have an understanding of what a node might look like.

A node layer is a row of those neuron-like switches that turn on and off as the input is fed through the net. Each layer's output is simultaneously the next layer's input, starting from an initial input layer receiving the data. Pairing the models' adjustable weights with input features is how we assign significance to those features with regard to how the neural network classifies and clusters input.

Neural networks with multiple hidden layers have each layer of node train on a distinct set of features based on the previous layer's output. More deep layers learn the more complicated features in the training data, since they are able to aggregate and recombine features from the previous layers. An example is shown in Figure 2-5.

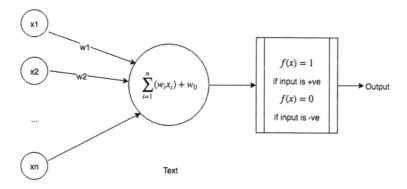

Figure 2-4. *Single node example*

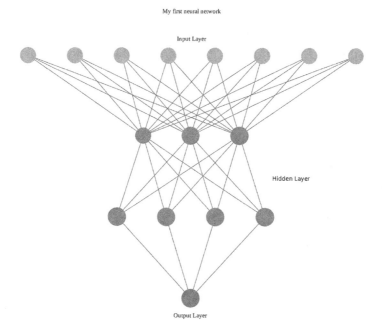

Figure 2-5. *Deep neural network*

2.8.8.1 Torch and tch-rs

Torch is a scientific computing framework with wide support for machine learning algorithms and has good support for GPU [8]. It is widely used for deep learning and creating neural network architectures. The C implementation of Torch is distributed in a package called `libtorch`. It provides a flexible N-dimensional array or tensor, which supports basic routines for indexing. slicing, transposing, type-casting, resizing, sharing storage, and cloning. This object is used by most other packages, and useful and more complicated routines are built on top of it.

In Rust, the `libtorch` C++-api has been extended using the tch-rs crate.[8] The aim of this create is to be as close to the C++ api as possible.

To use `tch-rs,` just adding `tch-rs` in your `Cargo.toml` and then doing a `cargo build` should work. If you are in a Mac system, you might encounter the error shown in Listing 2-54.

Listing 2-54. Possible error

```
dyld: Library not loaded: @rpath/libmklml.dylib
  Referenced from: /path to libtorch/lib/libcaffe2.dylib
  Reason: image not found
```

We can resolve this error by manually downloading the file and adding the paths to the lib file in the Mac terminal (Listing 2-55).

Listing 2-55. mkl installation

```
wget https://github.com/intel/mkl-dnn/releases/download/v0.18/
mklml_mac_2019.0.3.20190220.tgz
$ gunzip -c mklml_mac_2019.0.3.20190220.tgz| tar xvf -
$ export LD_LIBRARY_PATH=/path to mkl folder/lib:"$LD_LIBRARY_PATH"
```

---

[8]tch-rs.

The code should build after this.

We will now build a small neural network on it so that the network can train on the data. To be able to do that, we will need to change the data though. We will take a look at all the important steps next. These steps are under the assumption that all the data preprocessing steps are complete and are similar to the ones we encountered in the previous sections. The consolidated code is also present in the package iris_classification_tchrs for reference.

First, we will need to specify the test train ratio to be the same. This is a technicality that needs to be taken care because of the way matrix multiplication works and the implications that we will take a look at later (Listing 2-56).

Listing 2-56. chapter2/iris_classification_tchrs/src/simple_nn.rs

```rust
pub fn run() -> Result<(), Box<dyn Error>> {
  // previous code ...

  let test_size: f64 = 0.5;

  // remaining code ...
```

Now we will need to convert the vectors to torch tensors so that we are able to perform computations on those tensors (Listing 2-57).

Listing 2-57. chapter3/iris_classification_tchrs/src/main.rs

```rust
use tch::{kind, Kind, Tensor};

pub fn run() -> Result<(), Box<dyn Error>> {
  // previous code ...

  let flower_x_train = Tensor::float_vec(
    flower_x_train.as_slice());
  let flower_y_train = Tensor::float_vec(
    flower_y_train.as_slice()).to_kind(Kind::Int64);
```

```
let flower_x_test = Tensor::float_vec(
  flower_x_test.as_slice());
let flower_y_test = Tensor::float_vec(
  flower_y_test.as_slice()).to_kind(Kind::Int64);

// remaining code ...
```

We can now reshape the vectors to reflect the training size and the dimension of the features. Label values in this case are essentially vectors (Listing 2-58).

Listing 2-58. chapter3/iris_classification_tchrs/src/main.rs

```
pub fn run() -> Result<(), Box<dyn Error>> {
  // previous code ...

  let train_size = train_size as i64;
  let test_size = test_size as i64;
  let flower_x_train = flower_x_train.view(&[train_size, 4]);
  let flower_x_test = flower_x_test.view(&[test_size, 4]);
  let flower_y_train = flower_y_train.view(&[train_size]);
  let flower_y_test = flower_y_test.view(&[test_size]);

  // remaining code ...
```

Now we come to the actual neural network creation. Similar to Figure 2-4, let's create a single layer network where we are optimizing on the matrix equation $Y = X * W + B$ where all the values are vectors or matrices [9]. We will need to initiate our weights and biases for this (Listing 2-59).

Listing 2-59. chapter3/iris_classification_tchrs/src/main.rs

```
pub fn run() -> Result<(), Box<dyn Error>> {
  // previous code ...

  let mut ws = Tensor::ones(
    &[feature_length, 1], kind::FLOAT_CPU)
    .set_requires_grad(true);
  let mut bs = Tensor::ones(
    &[train_size], kind::FLOAT_CPU)
    .set_requires_grad(true);

  // remaining code ...
```

In the above example, we will be setting the requires_grad to be true because we will need to compute the gradients for these values. Tensors have requires_grad as false by default and gradient for tensors for which the requires_grad is false are not calculated [8].

Now we will need to first perform the matrix multiplication flower_x_ train * ws + bs. We will then consider the loss against flower_y_train. Then we need to propagate the loss. This operation will be repeated for multiple epochs, and we will report the accuracy for each epoch so that we are able to keep track of the increase or decrease in accuracy of the model (Listing 2-60).

Listing 2-60. chapter3/iris_classification_tchrs/src/simple_nn.rs

```
pub fn run() -> Result<(), Box<dyn Error>> {
  // previous part of the run code ...

  for epoch in 1..200 {
    let logits = flower_x_train.mm(&ws) + &bs;
    let loss = logits.squeeze().cross_entropy_for_logits(
      &flower_y_train);
    ws.zero_grad();
```

```
    bs.zero_grad();
    loss.backward();
    no_grad(|| {
      ws += ws.grad() * (-1);
      bs += bs.grad() * (-1);
    });
    let test_logits = flower_x_test.mm(&ws) + &bs;
    let test_accuracy = test_logits
      .argmax1(-1, false)
      .eq1(&flower_y_test)
      .to_kind(Kind::Float)
      .mean()
      .double_value(&[]);
    println!(
      "epoch: {:4} train loss: {:8.5} test acc: {:5.2}%",
      epoch,
      loss.double_value(&[]),
      100. * test_accuracy
    );
  }

  Ok(())
}
```

We expect to see an incremental decrease in loss of the model. The output should be similar to that in Listing 2-61.

Listing 2-61. chapter3/iris_classification_tchrs/src/simple_nn.rs

```
$ cd chapter2/iris_classification_tchrs
$ wget https://github.com/intel/mkl-dnn/releases/download/
v0.19/mklml_mac_2019.0.5.20190502.tgz
... other logs ...
```

```
Length: 28288113 (27M) [application/octet-stream]
Saving to: 'mklml_mac_2019.0.5.20190502.tgz'

mklml_mac_2019.0.5.20190502.tgz
100%[==========================================================
==============================>]  26.98M  1.07MB/s    in 2m 44s

2019-08-08 08:44:34 (169 KB/s) - 'mklml_mac_2019.0.5.20190502.
tgz' saved [28288113/28288113]

$ gunzip -c mklml_mac_2019.0.5.20190502.tgz| tar xvf -
x mklml_mac_2019.0.5.20190502/
...
x mklml_mac_2019.0.5.20190502/third-party-programs.txt
$ ls mklml_mac_2019.0.5.20190502/lib
libiomp5.dylib libmklml.dylib
$ export LD_LIBRARY_PATH=mklml_mac_2019.0.5.20190502/lib:"$LD_
LIBRARY_PATH"
$ cargo run nn < ../datasets/iris.csv
    Finished dev [unoptimized + debuginfo] target(s) in 0.05s
     Running `target/debug/iris_classification_tchrs nn`
Training data shape [300]
Training flower_y_train data shape [75]
epoch:    1 train loss:  1.69804 test acc: 30.67%
... printing for each epoch
epoch:  199 train loss:  1.10409 test acc: 29.33%
```

Linear network using Torch Now creating the previous model is great for understanding neural networks, but there is a better way of creating networks in torch similar to what is advocated for in pytorch as well. For creating models in torch, we can create a simple struct and then implement forward for the struct (Listing 2-62).

Listing 2-62. chapter2/iris_classification_tchrs/src/linear_with_
sgd.rs

```rust
use tch;
use tch::{nn, kind, Kind, Tensor, no_grad, vision, Device};
use tch::{nn::Module, nn::OptimizerConfig};

static FEATURE_DIM: i64 = 4;
static HIDDEN_NODES: i64 = 10;
static LABELS: i64 = 3;

#[derive(Debug)]
struct Net {
  fc1: nn::Linear,
  fc2: nn::Linear,
}

impl Net {
  fn new(vs: &nn::Path) -> Net {
    let fc1 = nn::Linear::new(vs,
      FEATURE_DIM, HIDDEN_NODES,
      Default::default());
    let fc2 = nn::Linear::new(vs,
      HIDDEN_NODES, LABELS,
      Default::default());
    Net { fc1, fc2 }
  }
}

impl Module for Net {
  fn forward(&self, xs: &Tensor) -> Tensor {
    xs.apply(&self.fc1).relu().apply(&self.fc2)
  }
}
```

In the above model, we create two linear networks, basically a hidden network between the input layer and the output labels. The forward method is then overriden to implement the network.

To train the above network, we will need to convert the Flower vectors to torch tensors (Listing 2-63).

Listing 2-63. chapter2/iris_classification_tchrs/src/linear_with_sgd.rs

```
pub fn run() -> Result<(), Box<dyn Error>> {
  // previous code with data loading and data splitting ...

  let flower_x_train = Tensor::float_vec(
    flower_x_train.as_slice());
  let flower_y_train = Tensor::float_vec(
    flower_y_train.as_slice()).to_kind(Kind::Int64);
  let flower_x_test = Tensor::float_vec(
    flower_x_test.as_slice());
  let flower_y_test = Tensor::float_vec(
    flower_y_test.as_slice()).to_kind(Kind::Int64);

  let train_size = train_size as i64;
  let test_size = test_size as i64;
  let flower_x_train = flower_x_train.view(
    &[train_size, FEATURE_DIM]);
  let flower_x_test = flower_x_test.view(
    &[test_size, FEATURE_DIM]);
  let flower_y_train = flower_y_train.view(
    &[train_size]);
  let flower_y_test = flower_y_test.view(
    &[test_size]);

  // remaining part of the fn ...
```

Now that the model and the appropriate tensors have been created, we should be able to train the model using an SGD algorithm (Listing 2-64).

Listing 2-64. chapter2/iris_classification_tchrs/src/simple_nn.rs

```
let vs = nn::VarStore::new(Device::Cpu); // use GPU for bigger
                                            models.
let net = Net::new(&vs.root());
let opt = nn::Adam::default().build(&vs, 1e-3)?;
for epoch in 1..200 {
  let loss = net
    .forward(&flower_x_train)
    .cross_entropy_for_logits(&flower_y_train);
  opt.backward_step(&loss);
  let test_accuracy = net
    .forward(&flower_x_test)
    .accuracy_for_logits(&flower_y_test);
  println!(
    "epoch: {:4} train loss: {:8.5} test acc: {:5.2}%",
    epoch,
    f64::from(&loss),
    100. * f64::from(&test_accuracy),
  );
};
```

We should be able to run the command `cargo run linear_with_sgd`
< `data/iris.csv` to run this linear model.

2.8.9 Model Evaluation

Similar to the regression section, in this classification section, we also have
been showing the accuracy of the models with the code for the models
so that the readers can execute the code and match the accuracy for
themselves. In this section the respective model evaluation functions that
we have used have been explained.

Accuracy The most common metric to evaluate a classification model is by using classification accuracy. It is the ratio of the number of correct predictions to the total number of input samples.

$$\text{Accuracy} = \frac{\text{Number of correct predictions}}{\text{Total number of predictions made}} \quad\quad (9)$$

We can implement this using the next function (Listing 2-65).

Listing 2-65. ml-utils/src/sup_metrics.rs

```
pub fn accuracy(y_test: &[u32], y_preds: &[u32]) -> f32 {
  let mut correct_hits = 0;
  for (predicted, actual) in y_preds.iter().zip(y_test.iter()) {
    if predicted == actual {
      correct_hits += 1;
    }
  }
  let acc: f32 = correct_hits as f32 / y_test.len() as f32;
  acc
}
```

The signature of the function is u32 because the y values will be specific labels.

Or since we are using rustlearn in this chapter, we can use the accuracy_score from the package (Listing 2-66).

Listing 2-66. chapter3/rustlearn_classification_tasks/src/logistic_reg.rs

```
use rustlearn::metrics::accuracy_score;

pub fn run() -> Result<(), Box<dyn Error>> {
  // previous code ...
```

```
let prediction = model.predict(&flower_x_test).unwrap();
let acc = accuracy_score(&flower_y_test, &prediction);

// remaining code ...
```

Using the above function, we should be able to see the following accuracy scores for the models implemented (Listing 2-67).

Listing 2-67. chapter3/rustlearn_classification_tasks/src/logistic_reg.rs}

```
$ cargo run lr < data/iris.csv
    Finished dev [unoptimized + debuginfo] target(s) in 3.66s
     Running `target/debug/rustlearn_classification_tasks lr`
Logistic Regression: accuracy: 0.3
Logistic Regression: accuracy: 0.3
```

This method of calculating the accuracy score only works if there are an equal number of samples belonging to each class. For example, let's assume that 99% of samples belong to class A and the remaining 1% belong to class B. Then by the above metric, a simple model, such as that in Listing 2-68, predicts all out-of-sample classes as class A would have an accuracy of 99%.

Listing 2-68. My awesome machine learning model

```
fn model(y_test: &Vec<f32>) -> Vec<String> {
  vec![String::from("Class A"); y_test.len()]
}
```

The result would be that we would have a false sense of achieving high accuracy.

Logarithmic Loss Logarithmic loss, or cross-entropy loss, works by penalizing the false classifications. It works well for multi-class classifications. Log Loss increases as the predicted probability diverges from the actual label. A perfect model would have a log loss of 0. So predicting a probability of 0.12 when the actual observation label is 1 would be bad and result in a high log loss [10]. This is given by

$$H_p(q) = -\frac{1}{N}\sum_{i=1}^{N} y_i \cdot \log(p(y_i)) + (1-y_i)\log(1-p(y_i)) \qquad (10)$$

The graph in Figure 2-6 shows the range of possible log loss given a true observation of 1. Log loss is not very steep when the probability is approaching 1 but increases rapidly when the probability is going toward 0. We want the behavior to be penalized for any errors, but notice how the penalizing is infinite when the predictors are confident and wrong.

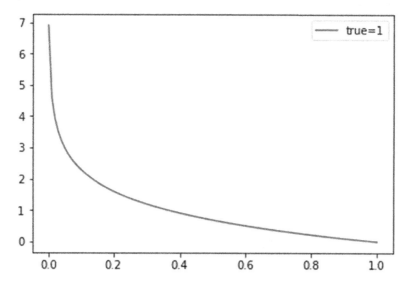

Figure 2-6. *Log loss*

Thus, log loss has a more nuanced approach to accuracy than a simple yes/no nature. It gives a value based on how wrong the model is from the true value.

In Rust, the function in Listing 2-69 will calculate the log-loss score. The vectors y_test and y_preds are the ground truths and the predicted values. Log loss is undefined for probability values of 0 and 1 and hence the y_test vector is clamped to a little above 0 and a little below 1 given by eps following the equation max(min(p, $1 - \in$), $\in$). Note that partial_cmp is used to compare as we are comparing between two floats. The Rust official stance is that comparing floating-point numbers is very tricky and situation dependent, and best avoided if at all possible. There is no panacea that "just works" [11].

Once done we will implement the actual log-loss function for binary classification on the clamped vector.

Listing 2-69. ml-utils/src/sup_metrics.rs

```rust
fn logloss_score(y_test: &Vec<f32>,
                 y_preds: &Vec<f32>,
                 eps: f32) -> f32 {
  let y_preds = y_preds.iter().map(|&p| {
    match p.partial_cmp(&(1.0 - eps)) {
      Some(Ordering::Less) => p,
      _ => 1.0 - eps, // if equal or greater.
    }
  });
  let y_preds = y_preds.map(|p| {
    match p.partial_cmp(&eps) {
        Some(Ordering::Less) => eps,
        _ => p,
    }
  });

  let logloss_vals = y_preds.zip(y_test.iter())
    .map(|(predicted, &actual)| {
      if actual as f32 == 1.0 {
```

```
      (-1.0) * predicted.ln()
    } else if actual as f32 == 0.0 {
      (-1.0) * (1.0 - predicted).ln()
    } else {
      panic!("Not supported. y_preds should be either 0 or 1");
    }
  });
  logloss_vals.sum()
}
```

This can now be used in something similar to Listing 2-70.

Listing 2-70. chapter2/rustlearn_classification_tasks/src/binary_
class_scores.rs

```
pub fn run() -> Result<(), Box<dyn Error>> {
  let preds = vec![1., 0.0001, 0.908047338626,
    0.0199900075962, 0.904058545833, 0.321508119045,
    0.657086320195];
  let actuals = vec![1., 0., 0., 1., 1., 0., 0.];
  println!("{:?}",
    logloss_score(&actuals, &preds, 1e-15)); // output 7.8581247
  Ok(())
}
```

ROC-AUC An ROC Curve (receiver operating characteristic curve)
is a graph showing the performance of a classification model at all
classification thresholds [12]. The curve plots two parameters.

- True Positive Rate
- False Positive Rate

True Positive Rate (TPR) is a synonym for recall and is therefore defined as

$$TPR = \frac{TP}{TP + FN} \qquad (11)$$

and False Positive Rate is defined as

$$FPR = \frac{FP}{FP + TN} \qquad (12)$$

An ROC Curve plots TPR vs. FPR at different classification thresholds. Lowering the classification threshold classifies more items as positive, thus increasing both false positives and true positives.

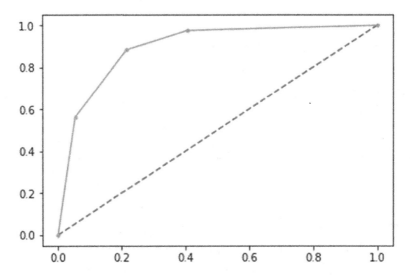

Figure 2-7. *Roc-auc curve*

AUC stands for "Area under the ROC Curve." It measures the entire two-dimensional area under the entire ROC Curve. For Figure 2-7 the ROC-AUC area is 0.895. As is evident, the score has a range between 0 and 1. The greater the value, the better is the performance of the model.

The crate `rustlearn` has roc-auc score implemented for binary classification (Listing 2-71).

Listing 2-71. chaper3/rustlearn_classification_tasks/src/binary_class_scores.rs

```
use rustlearn::metrics::roc_auc_score;

pub fn run() -> Result<(), Box<dyn Error>> {
  let preds = vec![
    1., 0.0001, 0.908047338626,
    0.0199900075962, 0.904058545833,
    0.321508119045, 0.657086320195];
  let actuals = vec![1., 0., 0., 1., 1., 0., 0.];
  println!("logloss score: {:?}",
    logloss_score(&actuals, &preds, 1e-15));
  println!("roc auc scores: {:?}",
    roc_auc_score(&Array::from(actuals), // new code
    &Array::from(preds))?);

  Ok(())
}
```

Running the above should give something like that shown in Listing 2-72.

Listing 2-72. chaper3/rustlearn_classification_tasks/src/binary_class_scores.rs

```
$ cargo run bs
    Finished dev [unoptimized + debuginfo] target(s) in 0.16s
     Running `target/debug/rustlearn_classification_tasks bs`
logloss score: 7.8581247
roc auc scores: 0.6666667
```

2.9 Conclusion

This chapter introduced you to different regression algorithms such as Linear Regression, Gaussian Processes, and Generalized Linear Models. Along with the algorithms, the package `rusty_machine` is introduced and how to create regression models of these algorithms using the package. Finally, we end the chapter with an understanding of how to evaluate regression models.

In the next chapter, you will learn about creating classification models.

2.10 Bibliography

[1]   *Introduction to Generalized Linear Models.* `https://newonlinecourses.science.psu.edu/stat504/node/216/`. 2018.

[2]   Heather Turner. *Introduction to Generalized Linear Models.* `http://statmath.wu.ac.at/courses/heather_turner/glmCourse_001.pdf`. 2018.

[3]   Alvira Swalin. *Choosing the Right Metric for Evaluating Machine Learning Models – Part I.* `https://medium.com/usf-msds/choosing-the-right-metric-for-machine-learning-models-part-1-a99d7d7414e4`. 2018.

[4]   Leo Breiman. *Random Forests.* `https://www.stat.berkeley.edu/~breiman/randomforest2001.pdf`. 2001.

[5]   Ramya Bhaskar Sundaram. *The math behind XGBoost.* `https://www.analyticsvidhya.com/blog/2018/09/an-end-to-end-guide-to-understand-the-math-behind-xgboost/`. 2018.

[6]   Tianqi Chen and Carlos Guestrin. "XGBoost: A
      Scalable Tree Boosting System." In: *Proceedings of
      the 22nd ACM SIGKDD International Conference on
      Knowledge Discovery and Data Mining.* KDD '16.
      ACM, 2016, pp. 785–794. ISBN: 978-1-4503-4232-2.
      DOI: 10.1145/2939672.2939785. url: `http://doi.
      acm.org/10.1145/2939672.2939785`.

[7]   Chih-Wei Hsu, Chih-Chung Chang, and Chih-Jen Lin.
      A Practical Guide to Support Vector Classification.
      `https://www.csie.ntu.edu.tw/~cjlin/papers/
      guide/guide.pdf`. Ed. Department of Computer
      Science. [Online; accessed 11-Nov-2019]. 2018.

[8]   *Torch.* `http://torch.ch/`.

[9]   *Excluding subgraphs from backward.* `https://
      pytorch.org/docs/stable/notes/autograd.
      html#excluding-subgraphs-from-backward`. 2018.

[10] *Log Loss.* `http://wiki.fast.ai/index.php/Log_
      Loss#Log_Loss_vs_Cross-Entropy`. 2017.

[11] *Float Comparisons.* `https://docs.rs/float-
      cmp/0.4.0/float_cmp/`.

[12] Google. *Classification: ROC Curve and AUC.*
      `https://developers.google.com/machine-
      learning/crash-course/classification/roc-
      and-auc`. 2019.

[13] Vitaly Bushaev. *Stochastic Gradient Descent with
      Momentum.* `https://towardsdatascience.com/
      stochastic-gradient-descent- with-momentum-
      a84097641a5d`. 2017.

[14]   Anastasios Kyrillidis. *Adagrad.* http://
akyrillidis.github.io/notes/AdaGrad.

[15]   Vitaly Bushaev. *L1 and L2 Regularization.*
https://towardsdatascience.com/l1-and-l2-
regularization-methods-ce25e7fc831c. 2017.

[16]   Maciej Kula. *Rustlearn Decision Tree.* https://
maciejkula.github.io/rustlearn/doc/rustlearn/
trees/decision_tree/index.html. 2018.

[17]   Rafael Irizarry. *Decision Tree.* https://rafalab.
github.io/pages/649/section-11.pdf. 2006.

[18]   *Decision Trees and Random Forest.* https://scikit-
learn.org/stable/modules/tree.html. 2011.

[19]   F. Pedregosa et al. "Scikit-learn: Machine Learning
in Python". In: *Journal of Machine Learning Research*
12 (2011), 2825–2830.

[20]   Leo Breiman and Adele Cutler. *Random Forests.*
https://www.stat.berkeley.edu/~breiman/
RandomForests/cc_home.htm.

[21]   StackExchange:user:15501. *Need help understanding
xgboost's approximate split points proposal.*
https://datascience.stackexchange.com/
questions/10997/need-help-understanding-
xgboosts-approximate- split-points-proposal.
2016.

[22]   *A Beginner's Guide to Neural Networks and Deep
Learning.* https://skymind.ai/wiki/neural-
network.

[23] *A Matrix Formulation of the Multiple Regression Model.* https://newonlinecourses.science.psu. edu/stat501/node/382/. 2018.

[24] Hsuan-Tien Lin and Chih-Jen Lin. *A Study on Sigmoid Kernels for SVM and the Training ofnon-PSD Kernels by SMO-type Methods.* https://www. csie.ntu.edu.tw/~cjlin/papers/tanh.pdf.

## CHAPTER 3

Unsupervised and Reinforcement Learning

In the previous chapters we took a look at the regression and classification algorithms that fall under the category of supervised algorithms [1]. In this chapter, we will be taking a look at the remaining forms of machine learning, namely unsupervised algorithms and reinforcement learning. In unsupervised algorithms, the labels or the target classes are not given. So the goal of unsupervised learning is to attempt to find natural partitions of patterns.

The main forms of performing unsupervised learning has been through clustering.

First, we will focus on different unsupervised learning algorithms and how we can implement them using Rust. We will be using the same Iris dataset, and the data preparation steps would be the same that we saw in Chapter 2. Finally, we should have two rusty_machine dense matrices `flower_x_train` and `flower_x_test`. Although this dataset is not totally conducive to unsupervised learning as we know the labels, this is a good dataset for understanding the workings of creating unsupervised models.

© Joydeep Bhattacharjee 2020
J. Bhattacharjee, *Practical Machine Learning with Rust*,
https://doi.org/10.1007/978-1-4842-5121-8_3

3.1 K-Means Clustering

A way of performing unsupervised learning is by observing which groups of data cluster together based on a notion of similarity. The simplest model that solves the clustering problem is k-means. The number of clusters needs to be set a priori; and based on that, the dataset is classified in such a way that these set numbers of clusters are formed. An example for a datasets clustered using k-means is shown in Figure 3-1.

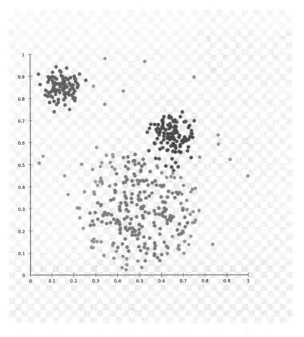

Figure 3-1. *K-means clustering: the number of clusters here is taken as 3, but we might as well have taken 2*

The output of the final k-means is quite dependent on the initial patterns that are used. The patterns are generally initialized using Forgy, Random Partition, or K-means++ methods. In the Forgy method, randomly chosen k observations from the dataset are used as the initial means. In the the random partition method, a cluster is assigned to each observation in the beginning and then proceeds to the updating step, thus computing the initial mean to be the centroid of the cluster's randomly assigned points. The Forgy method tends to spread the initial means out, while a Random partition places all of them close to the center of the dataset.

K-Means++ is almost the same as vanilla K-Means just that K-means++ starts with allocating one cluster center randomly and then searches for other centers given the first one [2].

Using rusty_machine we can use the K-MeansClassifier struct to implement and apply K-Means on the data using the train method. We can then either see the model centroids or run the predict method on the unseen data (Listing 3-1).

Listing 3-1. rust-machine-learning/chapter3/rusty_machine_ unsupervised/src/main.rs output

```
use rusty_machine as rm;
use rm::learning::k_means::KMeansClassifier;

fn main() -> Result<(), Box<Error>> {
  // data loading and splitting code ...

  const clusters: usize = 3;
  let model_type = "Kmeans";
  let mut model = KMeansClassifier::new(clusters);

  model.train(&flower_x_train)?;
  let centroids = model.centroids().as_ref().unwrap();
  println!("Model Centroids:\n{:.3}", centroids);

  println!("Predicting the samples...");
  let classes = model.predict(&flower_x_test).unwrap();
  println!("number of classes from kmeans: {:?}",
    classes.data().len());

  remaining code ...
```

The output for this code should be something like that shown in Listing 3-2.

Listing 3-2. rust-machine-learning/chapter3/rusty_machine_ unsupervised/src/main.rs output

```
$ cargo run < ../../chapter2/datasets/iris.csv
    Finished dev [unoptimized + debuginfo] target(s) in 0.04s
     Running `target/debug/rusty_machine_unsupervised`
Training the Kmeans model
Model Centroids:
    5.817 2.710 5.817 1.410
    4.988 3.388 4.988 0.260
    6.886 3.081 6.886 1.956
Predicting the samples...
number of classes from kmeans: 30
```

Creating this model initializes the k-means using k-means++. Apart from this, we can also use the Forgy or RandomPartition method (Listing 3-3).

Listing 3-3. rust-machine-learning/chapter3/rusty_machine_ unsupervised/src/main.rs output

```
use rm::learning::k_means::{KMeansClassifier, Forgy,
                            RandomPartition, KPlusPlus};

fn main() -> Result<(), Box<Error>> {
  // previous code ...

  // can use either Forgy or RandomPartition
  let mut model = KMeansClassifier::new_specified(3, 100, Forgy);

  //Train the model
  println!("Training the kmeans forgy model model");
  model.train(&flower_x_train)?;
```

```
let centroids = model.centroids().as_ref().unwrap();
println!("Model Centroids:\n{:.3}", centroids);

// Predict the classes and partition into
println!("Predicting the samples...");
let classes = model.predict(&flower_x_test).unwrap();
println!("number of classes from kmeans: {:?}", classes.
data().len());
println!("{:?}", classes.data().len());

// remaining code ...
}
```

Note that in the parameters to KMeansClassifier::new_specified we are passing 3 which is the number of partitions and we are specifying the number_of_epochs param at 100. The type if partition is Forgy in Listing 3-4, the output only for the code block in Listing 3-3 is shown.

Listing 3-4. rust-machine-learning/chapter3/rusty_machine_ unsupervised/src/main.rs output

```
Training the kmeans forgy model model
Model Centroids:
    5.018 3.414 5.018 0.264
    5.867 2.721 5.867 1.463
    6.873 3.045 6.873 1.976
Predicting the samples...
number of classes from kmeans: 30
30
```

3.2 Gaussian Mixture Model

A k-means algorithm is generally the first algorithm of choice when performing unsupervised learning, due to its relative simplicity in training and understanding. But the simple nature of k-means also brings in practical challenges when applied. In this section we will take a look at Gaussian Mixture Models (or GMM's), which are more generalized versions of k-means. GMM's give k-means a more theoretical and mathematical footing [3].

Think of the k-means as defining circle spheres (or a circle in the simplest sense) where if a point is within the circle, it's part of the cluster or the group; and if not within the circle, then it's not part of the group. The cluster center is the center of the circle and the radius can be the distance to the outermost point. In k-means this radius will act like a hard cutoff between which points are members of the cluster and which are not members of the cluster. In contrast, we can relax this hard boundary assumption and assume that all cluster centers have an effect on all points in the distribution. An interesting analogy in this respect might be that even though the planets in the solar system are part of the same cluster and revolve around the sun (an analogy for the cluster center), distant stars and galaxies do exert some gravitational pull on earth and the other planets. The pull though is quite minuscule so that for practical calculations we can ignore them.

Mathematically, this is achieved by assuming that all data points are generated from a mixture of a finite number of Gaussian distributions with unknown parameters. Hence, we go from a hard-bounded k-means to a more probabilistic model. A Gaussian distribution is show in Figure 3-2, and what the distribution looks like when we have multiple distributions is shown in Figure 3-3.

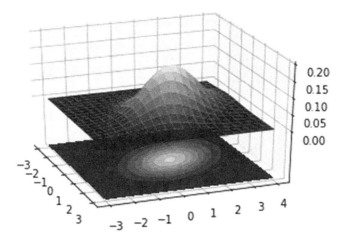

Figure 3-2. *Gaussian*

In Figure 3-3 we are using the Iris dataset to map the distribution between two variables: sepal_length and sepal_width.

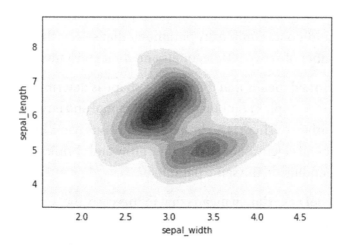

Figure 3-3. *Multiple distributions*

A Gaussian distribution is completely determined by its covariance matrix and its mean. The covariance matrix of a Gaussian distribution determines the directions and lengths of the aces of its density contours, all of which are ellipsoids. The different types of covariance matrices are the following:

- **Full** means that the components may independently adopt any position and shape.

- **Tied** means they have the same shape but the shape may be anything.

- **Diagonal** means that the contour axes are oriented along the coordinate axes, but otherwise the eccentricities may vary between components.

- **Spherical** means that the contour is a sphere.

- **Toeplitz** means that the diagonals of the contour have elements that share the same parameters. It has an additional complexity parameter that selects the number of active off-diagonals and their constants.

- **Shrinked** means that the contour shape is determined by the complex combination of a diagonal and any covariance. There is no immediate obvious tying, but it ties all eigenvalues so that they grow and shrink depending on the same parameter.

- **Kernel covariance** means that in this case, the covariance is defined by a positive definite function. We are in this case able to get a mixture model of functions (the greatest amount of generalization mathematically speaking).

Apart from the above, constraints can be put on the Gaussian mixture composition, which increases generalization of the model [4].

A Gaussian Mixture model is evaluated using the expectation maximization algorithm. This algorithm iterates to find the maximum likelihood estimates of parameters in statistical models. The advantage of using an EM algorithm is that it is able to estimate the parameters in models that depend on unobserved latent or hidden variables. In this case, we can think of the mixing probabilities of the Gaussian functions as the prior probabilities for the outcomes. For given values of individual components, the algorithm will evaluate the corresponding posterior probabilities, called responsibilities. These responsibilities are essentially the latent variables [5].

Rust Using rusty_machine, we can create a mixture model. Then we set the maximum number of iterations and the covariance types. We can then train this model using the train method. Once training is done apart from the predict method, we will also be able to get the means and the covariances of the trained model and run the predict method (Listing 3-5).

Listing 3-5. rust-machine-learning/chapter3/rusty_machine_unsupervised/src/main.rs

```
use rm::learning::gmm::{CovOption, GaussianMixtureModel};
use ml_utils::unsup_metrics::{jaccard_index, rand_index};

fn main() -> Result<(), Box<Error>> {
  // previous code ...

  let mut model = GaussianMixtureModel::new(2);
  model.set_max_iters(1000);
  model.cov_option = CovOption::Diagonal;

  println!("Training the model");
  model.train(&flower_x_train)?;

  // Print the means and covariances of the GMM
  println!("{:?}", model.means());
  println!("{:?}", model.covariances());
```

```
// Predict the classes and partition into
println!("Predicting the samples...");
let classes = model.predict(&flower_x_test).unwrap();
println!("number of classes from GMM: {:?}", classes.data().
len());

// Probabilities that each point comes from each Gaussian.
println!("number of Probablities from GMM: {:?}", classes.
data().len());

let predicted_clusters = flower_labels_clusters_gmm(classes.
data());
println!("predicted clusters from gmm: {:?}", predicted_
clusters);
println!("rand index: {:?}", rand_index(&predicted_clusters,
&flower_y_test_clus));
println!("jaccard index: {:?}", jaccard_index(&predicted_
clusters, &flower_y_test_clus));

// rest of the code ...
}
```

In place of Diagonal, we can also use the Full and Regularized options. Output for the above model is shown in Listing 3-6.

Listing 3-6. rust-machine-learning/chapter3/rusty_machine_ unsupervised/src/main.rs output

```
Training the Mixture model model
model means: Some(Matrix { rows: 3, cols: 4, data:
[5.00999871185655, 3.4275000972153493, 5.00999871185655,
0.24749396211890054, 6.661363633497823, 3.010315328597311,
6.661363633497823, 1.9381371828096539, 5.685719344467794,
2.670276005506212, 5.685719344467794, 1.3354911643349907] })
```

model covariances: Some([Matrix { rows: 4, cols: 4, data:
[0.13190228661798534, 0.0, 0.0, 0.0, 0.0, 0.16249683048609323,
0.0, 0.0, 0.0, 0.0, 0.13190228661798534, 0.0, 0.0, 0.0, 0.0,
0.011992091366866362] }, Matrix { rows: 4, cols: 4, data:
[0.25623508453320654, 0.0, 0.0, 0.0, 0.0, 0.06361683319359594,
0.0, 0.0, 0.0, 0.0, 0.25623508453320654, 0.0, 0.0, 0.0, 0.0,
0.11105964367994896] }, Matrix { rows: 4, cols: 4, data:
[0.14296761663516486, 0.0, 0.0, 0.0, 0.0, 0.06916983667682852,
0.0, 0.0, 0.0, 0.0, 0.14296761663516486, 0.0, 0.0, 0.0, 0.0,
0.06722138159035253] }])
Predicting the samples...
number of classes from GMM: 90
gmm classes: Matrix { rows: 30, cols: 3, data: [0.0000000000000
00000000000000000000000000000000000000000000000000000000001
4163053449105444, 0.9999999770633979, 0.0000000229366602131147763,
0.0000000000000000000000000000000000000000000000000000026043
664126255967, 0.055250462547009986, 0.94474953745299, 0.99999999
48172809, 0.000000000000230990802244913, 0.0000000051824961446
47463, 0.0000000000000000000000000000000000000000000636332801
3832893, 0.9996792118434199, 0.0003207881565800758, 0.9999987790
420889, 0.0000000208977320331112106, 0.0000120006017910087,
0.00000000000000000000000000000800914271537728, 0.0074920307322
04336, 0.9925079692677956, 0.0000000000000000000000000000000000
0000000000000000000000000000005285373201144257, 0.9998616296930777,
0.00013837030692238176, 0.9999995096444436, 0.0000000000415274
14529597685, 0.00000049035140337292042, 0.9999999132286986,
0.00000000000001393742090001966, 0.00000008676990757234786,
0.000000000000000000000000000000004855186479064118, 0.002719277
709391326, 0.9972807222906086, 0.0000000000000000000000000
0000000000000000000000000000000000000000000000000000000000000
0000000000000000000001355051993171768, 0.9999999999966558,

0.00000000000033441909895411436, 0.00000000000000000000000000000000
00000000005357830561231346, 0.9999562181964523, 0.0000437818035
47699205, 0.000000000000000000000000000000000000018392034391239705,
0.9999892831622388, 0.000010716837761224078, 0.0000000000000000
00000000003778865818805634, 0.7503202988312131, 0.2496797011687869,
0.00000000000000000000000000000000000000000000000000000000000000
00000000000000000000000000003893607802799448, 0.9999999999999958,
0.00000000000004298567960918699, 0.0000000000000130727701
77035594, 0.00023239823168584218, 0.999767601768301, 0.00000000
0000000000000000000000000000000000000015160002176660837,
0.8387135667053368, 0.16128643329466325, 0.00000000000000000000
00000000000000000000000000000000000000000000000000000000000000000
000000000034953327151517393, 0.9999997656317542, 0.0000002343682
4583147217, 0.000000000000000000000002340633434163898, 0.00050881
61744897154, 0.9994911838255103, 0.000000000000000000000000000000
00000000000000000000000000000000000000000000689099149495584, 0.99999
99977229123, 0.0000000022770877848717026, 0.9999999436031238,
0.000000000005926988070694233, 0.0000000056390949161758715,
0.9999999132286986, 0.0000000000001393742090001966, 0.0000000867
6990757234786, 0.0000000000000000000000000000000000000000000000
00000000000000000000000000000000005362212435215233, 0.9999122879
769339, 0.00008771202306615775, 0.9999997736979629, 0.000000000
03757036119192523, 0.000002262644667433647, 0.0000000000000000
000000000000000034483840914121184, 0.9985361511608948, 0.001463
8488391051328, 0.00000000000000000000000000000000000000000000
004937468317251526, 0.42540236377099694, 0.5745976362290031,
0.9999999948172809, 0.0000000000002230990802244913, 0.000000005
182496144647463, 0.9999998751000135, 0.0000000000421445815
8918603, 0.0000012485784186635827, 0.00000000000000000000000000
0000000025705463248504494, 0.262309436768884, 0.7376905632311159,
0.9999999956801614, 0.0000000000012481190923957532,
0.0000000043197138843938396] }

```
number of Probablities from GMM: 90
predicted clusters from gmm: [{21, 23, 2, 8, 26, 4, 7, 20, 27,
29}, {13, 3, 10, 14, 16, 6, 22, 24, 0, 12, 17, 11, 19}, {15, 9,
1, 5, 18, 25, 28}]
rand index: 0.7793103448275862
jaccard index: 0.48936170212765956
```

3.3 Density-Based Spatial Clustering of Applications with Noise (DBSCAN)

The idea behind DBSCAN is that clusters are essentially regions of high density separated by areas of low density. Much like cities, which are generally highly densely interspaced with the countryside, which are of lower density. High density means it's probably a cluster and low density is assumed to be noise. An interesting difference with k-means is that k-means assumes that the underlying clusters are convex shaped, but that assumption is not necessary for discovering clusters using DBSCAN; hence they are potentially more general. Thus, this algorithm is particularly useful when it seems like it's a large dataset with complicated shapes for the clusters and lots of noise in the dataset.

From an evaluation point of view, each point P is evaluated and there must be at least a certain number of points m_1, m_2,... within a set radius, R of the point P. This is the minimum density that the cluster must have. The algorithm needs three input parameters [6].

- k, the nearest neighbor list size;

- eps, the radius that delimits the neighborhood area of a point;

- min points, the minimum number of points that must exist in the eps neighborhood.

The issue with DBSCAN is that it cannot handle varying densities. Also, this algorithm is quite sensitive to parameters set.

DBSCAN models can be created using rusty_machine. In Listing 3-7 we create a DBSCAN model with eps as 3 and minimum samples as 10. We then pass true flag to set_predict method of the model. This allows us to use the predict method on new unseen data. Similar to previous models, we can train and predict on the dataset. Apart from these, we can also check the clusters that the model has learned.

Listing 3-7. rusty_machine_unsupervised

```
use rm::learning::dbscan::DBSCAN;
use rm::learning::UnSupModel;

fn main() -> Result<(), Box<Error>> {
  // previous code ...

  let mut model = DBSCAN::new(0.3, 10);
  model.set_predictive(true);
  model.train(&flower_x_train)?;

  let clustering = model.clusters().unwrap();
  let classes = model.predict(&flower_x_test).unwrap();

  // remaining code ...
```

Apart from DBSCAN::new we can use DBSCAN::default. The default values that are initialized are 0.5 for eps, 5 for minimum points.

3.4 Principal Component Analysis

One of the main components of machine learning is matrix multiplication. Matrix multiplications are generally computationally expensive [7]. Also, the number of dimensions that we are dealing with needs to be taken into account, and we should not add too many dimensions unnecessarily. We are shown by the Hughes Phenomenon, as seen in Figure 3-4, that as the number of features increases, a classifiers performance increases as well until we reach the optimal number of features. Adding more features for the same size as the training set will degrade the features. This is called the curse of dimensionality [8].

Many algorithms such as KNN are particularly susceptible to this curse of dimensionality. A way to escape this curse is by dimensionality reduction. In dimensionality reduction, we generally choose a mathematical representation within which most of the variance in the original data, if not all, can be explained. The effect is that we are able to remove a significant number of features while retaining a lot of the information.

Principal Component Analysis is a method of dimensionality reduction. It's essentially a transformation where our original variables will get converted to a new set of variables, which are linear combinations of the original set of variables. It means that if the original features are "a" and "b," for example, the resultant features might be $c = p_1a + q_1b$ and $d = p_2a + q_2b$.

$$PX = Y \qquad (1)$$

where X is the original recorded dataset, Y is the representation of the dataset, and P is the linear transformation matrix. Geometrically, we can see that P is a rotation and stretches from X to Y.

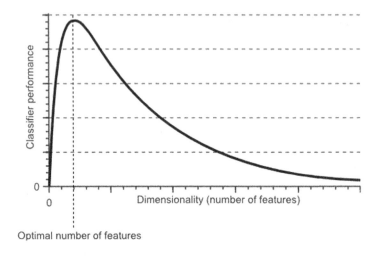

Optimal number of features

Figure 3-4. *Hughes Phenomenon*

PCA and Rust Although PCA has been implemented in `rusty_machine`, it has not been published in the crate as of the writing of this book. Hence, we will need to update the dependencies in Cargo.toml file to pull in the latest code from the master (Listing 3-8).

Listing 3-8. Cargo.toml

```
[dependencies]
rusty-machine = { git = "https://github.com/AtheMathmo/rusty-machine.git", rev = "f43da8d" }
```

Now we should be able to create a PCA model and train on the data. In this case we are reducing the dimensions to 2 and asking the model to center the clusters (Listing 3-9).

Listing 3-9. rusty_machine_unsupervised

```
use rm::learning::pca::PCA;
use rm::learning::UnSupModel;
let mut model = PCA::new(2, true);
model.train(&flower_x_train)?;
```

```
println!("{:?}", model.predict(&flower_x_test)?);
println!("{:?}", model.components());
```

We can also use the `PCA::default` method to create a default model, but the default model has all the components. So, we will be having a reduction in the dimensions.

3.5 Testing an Unsupervised Model

Evaluating the performance of an unsupervised model is difficult as there are no labels to compare the final score with. One way is through internal metrics such as the silhouette score, which aims at formalizing the attainment of high intra-cluster similarity; or points within a cluster should be close to each other and have low inter-cluster similarity, which means similarity between points in two clusters should be low. But good scores on an internal criterion may not necessarily translate into good effectiveness in an application. The other approach is through direct evaluation in the application of interest. For example, a website implementing search may measure the time taken by the users to find an answer with different clustering algorithms, and the two algorithms can be compared with beta testing. This is the most direct evaluation, but it is expensive, especially if large user studies are necessary.

A third approach is by using a surrogate of user judgments, in which case we use a set of classes and create a gold standard ourselves. The gold standard is ideally produced by human judges with good levels of inter-judge agreement. We can then compute an external criterion that evaluates how well the cluster matches the gold standard classes. In this section, two measures of external criteria are described with code accompanying them.

Rand Index The Rand Index computes a similarity measure between two clusters by considering all pairs of samples and counting pairs that are assigned in the same or different clusters in the predicted and true clustering. The most common formulation of the Rand Index focuses on the following four sets of different permutations given by $\binom{n}{k}$ element pairs:: N_{11} is the number of element pairs that are grouped in the same cluster in both clustering, N_{10} is the number of element pairs that are grouped in the same cluster by A but in different clusters by B, N_{01} is the number of element pairs that are grouped in the same cluster by B but in different clusters by A, and N_{00} is the number of elements pairs that are grouped in different clusters by both A and B. Notice that N_{11} and N_{00} are the indicators of agreements between clusters A and B while N_{10} and N_{01} are the disagreements.

Table 3-1. *Contingency Table*

A/B	B_1	B_2	$\ldots$	B_n	Sums
A_1	n_{11}	n_{12}	$\ldots$	n_{1n}	a_1
A_2	n_{21}	n_{22}	$\ldots$	n_{2n}	a_2
$\vdots$	$\vdots$	$\vdots$	$\ddots$	$\vdots$	$\vdots$
A_n	n_{n1}	n_{n2}	$\ldots$	n_{nn}	a_n
Sums	b_1	b_2	$\ldots$	b_n	N

The contingency is shown in Table 3-1. Therefore, the Rand index would be given by the below function:

$$RI(A, B) = \frac{N_{11} + N_{00}}{\binom{n}{k}} \tag{2}$$

The value of the above index would lie between 0 and 1, where 1 indicates that the clusterings are identical and 0 means that the clusters do not share a single pair of elements.

Rand index is implemented in ml-utils (Listing 3-10).

Listing 3-10. ml-utils/src/unsup_metrics.rs

```
pub fn rand_index(clusters1: &[HashSet<u64>],
    clusters2: &[HashSet<u64>]) -> f64 {
  let (n11, n10, n01, n00) = count_pairwise_cooccurence(
    clusters1, clusters2);
  (n11 + n00) / (n11 + n10 + n01 + n00)
}
```

The implementation of a count_pairwise_cooccurence function can be found on the same module but has been skipped for brevity.

We should now be able to run this function and get the index (Listing 3-11).

Listing 3-11. ml-utils/src/unsup_metrics.rs

```
use ml_utils::unsup_metrics::rand_index;

println!("rand index: {:?}",
  rand_index(&predicted_clusters, &flower_y_test_clus));
```

Jaccard Index Similar to the Rand index, we have the Jaccard Index. It is defined by the size of the intersection divided by the size of the union.

$$J(A,B) = \frac{|A \cap B|}{|A \cup B|} \qquad (3)$$

This can be easily understood by Figure 3-5. The intersection should approach the union as the clusters are similar to each other.

Take a look at the implementation of the Jaccard index in ml-utils (Listing 3-12).

Listing 3-12. ml-utils/src/unsup_metrics.rs

```
pub fn jaccard_index(clusters1: &[HashSet<u64>],
    clusters2: &[HashSet<u64>]) -> f64 {
  let (n11, n10, n01, n00) = count_pairwise_
  cooccurence(clusters1, clusters2);
  let denominator = n11 + n10 + n01;
  if denominator > 0.0 {
    return n11 / denominator;
  } else {
    0.0
  }
}
```

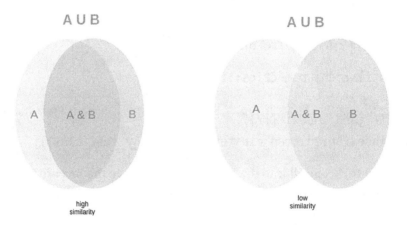

Figure 3-5. *Jaccard*

We should now be able to run this function and get the index (Listing 3-13).

Listing 3-13. ml-utils/src/unsup_metrics.rs

```
use ml_utils::unsup_metrics::jaccard_index;

println!("jaccard index: {:?}",
  jaccard_index(&predicted_clusters, &flower_y_test_clus));
```

3.6 Reinforcement Learning

Reinforcement learning is the utilization of different algorithms so that a suitable action can be chosen to maximize rewards in a given situation. The difference from supervised learning is that the labeled input/output pairs may not be present. In unsupervised learning, we are interested in finding the similarities and differences between the data points. But reinforcement learning is more about finding the best behavior given an environment. To build a reinforcement system, the software system should have a mechanism to make observations and take actions within an environment. In return, it should receive rewards in some form. The idea is to maximize long-term rewards.

There are various real-world applications for reinforcement learning. Robots have been used in manufacturing for some time now. Some repetitive tasks such as picking an object and putting it in a container benefit greatly from using reinforcement learning in their programming. Whether they succeed or fail, they are able to train themselves and complete the tasks with good speed and accuracy. Another field where there are good applications of reinforcement learning is speculative trading in the stock market. Agents are being used to pick between trading strategies using Q-learning, using one simple instruction to maximize the value of the portfolio. Reinforcement learning is of great value in such systems that are highly controlled and which require speed to choose between a high range of action.

The algorithm that is used by the software to determine its actions is called its policy. The policy should have access to two functions of the target body, a way to take observations as inputs and a way to take the next step.

One of the challenges of reinforcement learning is that in order to train an agent, we need to create the environment. The environment can be both real-world and software simulations. Real-world simulations are outside the scope of this book so we will target a subset of software simulations. One of the great crates for reinforcement learning is rsrl. Since the api is not stable yet, we will use a slightly modified version of the code to be able to call the functions well (Listing 3-14). Hence we have the ndarray crate for matrix manipulation and slog for logging. The crate to be focused on in this section is rsrl which will provide us with the relevant types for reinforcement learning.

Listing 3-14. chapter3/rsrl_custom/Cargo.toml

```
[package]
name = "rsrl_custom"
version = "0.1.0"
edition = "2018"

[dependencies]
rsrl = { git = "https://github.com/infinite-Joy/rsrl", branch =
"mymodel" }
slog = "2.4.1"
ndarray = "0.12.0"
```

Now we should be able to create a custom model. Most of the code here has been taken from the A2C examples and the Cart-pole problem in reinforcement learning. The Cart-pole problem, also known as the inverted pendulum, is a pendulum with the center of gravity above the pivot point. This results in the pendulum being very unstable. The goal is to keep the pole balanced by applying forces in the appropriate direction at the pivot point. This is shown in Figure 3-6.

Figure 3-6. *Cart-pole*

At each time step, we are able to observe its position *x*, velocity *dx*, angle *θ*, and angular velocity *dθ*. Thus, the state space has four dimensions of continuous values and the action space has one dimension of two discrete values. The force can be represented by ALL_ACTIONS. (Listing 3-15).

Listing 3-15. chapter3/rsrl_custom/src/main.rs

```
const ALL_ACTIONS: [f64; 2] = [-1.0 * CART_FORCE,
1.0 * CART_FORCE];
```

And we will need to define the environment by the scope of the current state and what will happen when a force is applied (Listing 3-16).

Listing 3-16. chapter3/rsrl_custom/src/main.toml

```
impl CartPole {
  fn new(x: f64, dx: f64, theta: f64, dtheta: f64) -> CartPole {
    CartPole {
      state: Vector::from_vec(vec![x, dx, theta, dtheta]),
    }
  }
}

  fn update_state(&mut self, a: usize) {
    let fx = |_x, y| CartPole::grad(
      ALL_ACTIONS[a], &y); // when a force <a> is applied
```

```
  let mut ns = runge_kutta4(
    &fx, 0.0, self.state.clone(), TAU);

  ns[StateIndex::X] = clip!(
    LIMITS_X.0, ns[StateIndex::X], LIMITS_X.1);
  ns[StateIndex::DX] = clip!(
    LIMITS_DX.0, ns[StateIndex::DX], LIMITS_DX.1);

  ns[StateIndex::THETA] = clip!(
    LIMITS_THETA.0, ns[StateIndex::THETA], LIMITS_THETA.1);
  ns[StateIndex::DTHETA] = clip!(
    LIMITS_DTHETA.0, ns[StateIndex::DTHETA], LIMITS_DTHETA.1);

  self.state = ns;
}

fn grad(force: f64, state: &Vector) -> Vector {
  let dx = state[StateIndex::DX];
  let theta = state[StateIndex::THETA];
  let dtheta = state[StateIndex::DTHETA];

  let cos_theta = theta.cos();
  let sin_theta = theta.sin();

  let z = (force
          + POLE_MOMENT
          * dtheta * dtheta
          * sin_theta) / TOTAL_MASS;

  let numer = G * sin_theta - cos_theta * z;
  let denom = FOUR_THIRDS
    * POLE_COM
    - POLE_MOMENT
    * cos_theta * cos_theta;
```

```
    let ddtheta = numer / denom;
    let ddx = z - POLE_COM * ddtheta * cos_theta;

    Vector::from_vec(vec![dx, ddx, dtheta, ddtheta])
  }
}
```

In rsrl, to implement this environment, we will need to implement
the Domain trait. If we see the signature of the Domain trait, we get an
understanding of the types of behavior that we need to implement. This
can be seen in https://github.com/tspooner/rsrl/blob/master/src/
domains/mod.rs

```
pub trait Domain {
    type StateSpace: Space;
    type ActionSpace: Space;
    fn emit(&self) -> .. // this is for observation
    fn step( .. // given an action what should be the next state.
    fn is_terminal(&self) -> bool; // this is to end the sequence
                                    of action sometime.
    fn reward( .. // a reward mechanism
    fn state_space(.. // returns an instance of state space
    fn action_space(.. // returns an instance of the action
                        class.
}
```

Coming back to Cart-pole, first we define a default (Listing 3-17).

Listing 3-17. chapter3/rsrl_custom/src/main.rs

```
impl Default for CartPole {
  fn default() -> CartPole { CartPole::new(0.0, 0.0, 0.0, 0.0)
}
}
```

Now we should be able to implement the Domain behaviors for Cart-pole. We first define the state space and the action space. We will choose the state space to be a LinearSpace, which is an n dimensional homogeneous space as defined in the spaces repo.[1] The action space will need to be an Ordinal, which is the type for defining a finite, ordinal set of values in spaces packages (Listing 3-18).

Listing 3-18. chapter3/rsrl_custom/src/main.rs

```
use rsrl::geometry::product::LinearSpace;
use rsrl::geometry::discrete::Ordinal;

impl Domain for CartPole {
  type StateSpace = LinearSpace<Interval>;
  type ActionSpace = Ordinal;
```

Emit is simple. If it's the last state, then we define it as terminal or else we define it as full (Listing 3-19).

Listing 3-19. chapter3/rsrl_custom/src/main.rs

```
impl Domain for CartPole {
  // code ..

  fn emit(&self) -> Observation<Vector<f64>> {
    if self.is_terminal() {
      Observation::Terminal(self.state.clone())
    } else {
      Observation::Full(self.state.clone())
    }
  }
}
```

---

[1] https://github.com/tspooner/spaces.

```
fn is_terminal(&self) -> bool {
  let x = self.state[StateIndex::X];
  let theta = self.state[StateIndex::THETA];

  x <= LIMITS_X.0
    || x >= LIMITS_X.1
    || theta <= LIMITS_THETA.0
    || theta >= LIMITS_THETA.1
}
```

Now we will need to implement behaviors for reward and state_
space. The reward behavior will be the final or terminal reward when the
observation is the final observation in the space of possible observations.
For other sets of observations, the reward should be the reward step
(Listing 3-20).

Listing 3-20. chapter3/rsrl_custom/src/main.rs

```
impl Domain for CartPole {
 // previous method definitions

 fn reward(&self, _: &Observation<Vector<f64>>, to:
 &Observation<Vector<f64>>) -> f64 {
   match *to {
     Observation::Terminal(_) => REWARD_TERMINAL,
     _ => REWARD_STEP,
   }
 }
}

// remaining methods.
```

A state space behavior is within the range of the possible behaviors that
our agent can have or, in other words, the constraints that our environment
will have on any agent. Since in our case we are trying to model a Cart-pole,

the state space will be a combination of the empty space and bounded values for position x, velocity dx, angle θ, and angular velocity $d\theta$ as defined previously (Listing 3-21).

Listing 3-21. chapter3/rsrl_custom/src/main.rs

```
impl Domain for CartPole {
  // previous method definitions ...

  fn state_space(&self) -> Self::StateSpace {
    LinearSpace::empty() + Interval::bounded(LIMITS_X.0,
    LIMITS_X.1)
      + Interval::bounded(LIMITS_DX.0, LIMITS_DX.1)
      + Interval::bounded(LIMITS_THETA.0, LIMITS_THETA.1)
      + Interval::bounded(LIMITS_DTHETA.0, LIMITS_DTHETA.1)
  }

  // remaining methods ...
```

And the action space will be just the value 2 signifying that some dimension of the state space will be changed by 2 on application of action, and that is the only possible difference in any two successive states (Listing 3-22).

Listing 3-22. chapter3/rsrl_custom/src/main.rs

```
impl Domain for CartPole {
  // previous method definitions ...

  fn action_space(&self) -> Ordinal { Ordinal::new(2) }
}
```

Currently there are different reinforcement algorithms that can be used to train the environment. We have the actor critic system, doing this using Q-learning, with some variants. In the example, we have used the actor critic system. In this, there are essentially two systems that need to be trained. One is the critic that measures how good the action taken is and the actor that controls how the agent behaves. So, we will need to first define the agent and the critic. To define both the agent and the critic, we will first define a policy. A policy is what the agent will need to do to accomplish the task at hand. Keep in mind that the goal of reinforcement learning is to arrive at the optimal policy, and hence it is quite important that the policy is chosen carefully. In this case we can define the policy to be a Gibbs kernel sampler of a vector composed of linear function approximators from the LFA library.[2] The LFA library provides the type LFA, which is a framework for linear function approximation with gradient descent. The advantage of using a linear function of features in the policy is that now we can approximate the Q-function using SARSA[0] and assign it to be the critic. Further. the agent will be derived as an A2C type from both the critic and the policy (Listing 3-23).

Listing 3-23. chapter3/rsrl_custom/src/main.rs

```
use rsrl::fa::fixed::Fourier;
use rsrl::fa::LFA;
use rsrl::policies::parameterised::Gibbs;
use rsrl::control::td::SARSA;
use rsrl::control::actor_critic::A2C;

fn main() {
  let domain = CartPole::default();

  let n_actions = domain.action_space().card().into();
  let bases = Fourier::from_space(
    3, domain.state_space()).with_constant();
```

---

[2]https://github.com/tspooner/lfa.

```
let policy = make_shared({
  let fa = LFA::vector(bases.clone(), n_actions);

  Gibbs::new(fa)
});
let critic = {
  let q_func = LFA::vector(bases, n_actions);

  SARSA::new(q_func, policy.clone(), 0.1, 0.99)
};

let mut agent = A2C::new(critic, policy, 0.01);
// rest of the code ..
```

Once they are created, we can go ahead and train the system to find the optimal policy. In this case we can go through 1,000 episodes to get the optimal policy (Listing 3-24).

Listing 3-24. chapter3/rsrl_custom/src/main.rs

```
fn main() {
  // previous code ...

  let logger = logging::root(logging::stdout());
  let domain_builder = Box::new(CartPole::default);

  let _training_result = {
    let e = SerialExperiment::new(&mut agent, domain_builder.
    clone(), 1000);
    run(e, 1000, Some(logger.clone()))
  };

  let testing_result = Evaluation::new(
    &mut agent, domain_builder).next().unwrap();

  info!(logger, "solution"; testing_result);
}
```

Running this would return the number of steps that are required and the optimal policy. Since the `testing_result` is a result of the `Evaluation` type, we would need the logger to print it out.

3.7 Conclusion

In this chapter we explored unsupervised learning and reinforcement learning. In unsupervised learning, we looked at how to implement K-means, Gaussian mixture models, and DBSCAN for clustering and PCA for dimensionality reduction using rusty_machine using a dataset without labels. Then we looked at evaluation clustering models using two popular methods: the Rand index and the Jaccard index. The basic methods of implementing other evaluation criteria would be similar.

Last, we explored reinforcement learning and how to implement a custom model using it in rsrl.

In the next chapter we will be moving away from models and focusing on another important dimension of machine learning, which is how to work with common data formats and transformation techniques. We will also learn how to create structured data if the data available is not structured at all.

3.8 Bibliography

[1]   Zoubin Ghahramani. *Unsupervised Learning.*
      `http://mlg.eng.cam.ac.uk/pub/pdf/Gha03a.pdf`.
      2004.

[2]   David Arthur and Sergei Vassilvitskii. *k-means++:*
      The Advantages of Careful Seeding. `http://ilpubs.`
      `stanford.edu:8090/778/1/2006-13.pdf`. 2006.

[3]   Jake VanderPlas. "Python DataScience Handbook. Essential tools for working with data." In: O'Reilly Media, 2016. Chap. In Depth: Gaussian Mixture Models.

[4]   Jugurta Montalv ao Janio Canuto, ed. *CONTOUR LEVEL ESTIMATION FROM GAUSSIAN MIXTURE MODELSAPPLIED TO NONLINEAR BSS* (Brazil). Universidade Federal de Sergipe (UFS)N ucleo de Engenharia El etricaS ao Crist ov ao, 2008.

[5]   *The EM Algorithm for Gaussian Mixtures.* https://www.ics.uci.edu/~smyth/courses/cs274/notes/EMnotes.pdf.

[6]   T. Soni Madhulatha. *An Overview on Clustering Methods.* https://arxiv.org/abs/1205.1117. 2012.

[7]   Joydeep Bhattacharjee. *Dimensionality Reduction and Principal Component Analysis - I.* https://medium.com/technology-nineleaps/dimensional-reduction-and-principal-component-analysis-i-8ce60a5ed2c2. 2017.

[8]   Badreesh Shetty. *Curse of Dimensionality.* https://towardsdatascience.com/curse-of-dimensionality-2092410f3d27. 2019.

[9]   Hastie and Tibshirani. *Gaussian Mixture Models.* http://statweb.stanford.edu/~tibs/stat315a/LECTURES/em.pdf. 2008.

[10] Lars Kai Hansen Rasmus Elsborg Madsen and Ole Winther. http://www2.imm.dtu.dk/pubdb/views/edoc_download.php/4000/pdf/imm4000.

[11]   Yong-Yeol Ahn Alexander J. Gates. *The Impact of Random Models on Clustering Similarity.* http://www.jmlr.org/papers/volume18/17-039/17-039.pdf. 2017.

[12]   David Poole and Alan Mackworth. *SARSA with Linear Function Approximation.* https://artint.info/html/ArtInt_272.html. 2010.

CHAPTER 4

Working with Data

Machine learning applications run on data. One of the main activities when building a machine learning application is to collate from different data sources, store in an effective format, and transform of raw data into formats that are appropriate for the machine learning app. Data can come in different formats. In the previous chapters, we mainly worked with CSV files. CSV files are great for storing and retrieving information. There is the added advantage that data paradigms in the CSV translate quite well to matrix formats where most of the calculations happen. But in reality, when working on actual data, we seldom find that the data is presented in a nice CSV format. In this chapter we will explore the different types of popular paradigms to store and retrieve information and see how we can leverage them to extract, process, and store data.

4.1 JSON

JSON stands for JavaScript Object Notation. It is a very common format used for asynchronous browser-server communication; and hence, a lot of time the data from some web process would be stored directly in the JSON format. Many developers consider JSON to be the default serialization-deserialization data structure for the web, the result being that many web apis publish their data in the JSON format. In this section we will see how we can work with JSON data.

© Joydeep Bhattacharjee 2020
J. Bhattacharjee, *Practical Machine Learning with Rust*,
https://doi.org/10.1007/978-1-4842-5121-8_4

Serde and related crates are great for serialization and deserialization of different types of data formats. For JSON we have the serde_json.

To have a look at how to work with JSON data structure and create code for parsing JSON objects to Rust data types, let us create a package named data_formats (Listing 4-1).

Listing 4-1. bin package creation for data_formats

```
$ cd chapter4 && cargo new data_formats -bin
```

Now inside the folder data_formats, open the Cargo.toml file and update the contents with the dependencies shown in Listing 4-2. As we have seen before serde and serde-derive is for the data management and the serde_json will be for parsing the json files.

Listing 4-2. chapter4/working_with_data/data_formats/Cargo.toml

```
[package]
name = "data_formats"
version = "0.1.0"
edition = "2018"

[dependencies]
serde = "1.0.90"
serde_derive = "1.0.90"
serde_json = "1.0"
```

Now we should be able to read an arbitrary JSON string and get the structure of the JSON. This is enabled by the serde_json::Value ENUM (Listing 4-3).

Listing 4-3. chapter4/working_with_data/data_formats/src/
jsonreading.rs

```
pub fn run() -> Result<(), Box<dyn Error>> {
  let json_str = r#"{
    "FirstName": "John",
    "LastName": "Doe",
    "Age": 43,
    "Address": {
      "Street": "Downing Street 10",
      "City": "London",
      "Country": "Great Britain"
    },
    "PhoneNumbers": [
      "+44 1234567",
      "+44 2345678"
    ]
  }"#;
  let person: serde_json::Value = serde_json::from_str(json_str)
    .expect("JSON was not well-formatted");
  let address = person.get("Address").unwrap();
  println!("{:?}", address.get("City").unwrap());
}
```

To be able to run the code, we will need to call the function in the
main function. Similar to other packages that we have seen so far, we can
define the main.rs file containing the main method, which will only parse
the incoming arguments. If the incoming argument is json, then we will
pass the control to the run function in the jsonreading module. For that,
ofcourse, we will need to expose the jsonreading module using the mod
keyword (Listing 4-4).

Listing 4-4. chapter4/working_with_data/data_formats/src/main.rs

```rust
#[macro_use]
extern crate serde_derive;

extern crate serde;
extern crate serde_json;
extern crate serde_xml_rs;

use std::vec::Vec;
use std::process::exit;
use std::env::args;

mod jsonreading;
mod xmlreading;
mod csvreading;

fn main() {
    let args: Vec<String> = args().collect();
    let model = if args.len() < 2 {
        None
    } else {
        Some(args[1].as_str())
    };
    let res = match model {
        None => {println!("nothing",); Ok(())},
        Some("json") => jsonreading::run(),
    };
    // Putting the main code in another function serves two
       purposes:
    // 1.We can use the '?' operator.
    // 2.We can call exit safely, which does not run any
          destructors.
    exit(match res {
```

```
    Ok(_) => 0 ,
    Err(e) => {
      println!("{}", e);
      1
    }
  })
}
```

Now that we have the constructs in place, we can run the package and get the output (Listing 4-5).

Listing 4-5. chapter4/working_with_data/data_formats/src/jsonreading.rs

```
$ cargo run json
    Finished dev [unoptimized + debuginfo] target(s) in 0.05s
     Running 'target/debug/data_formats json'
String("London")
```

Although the code in Listing 4-3 works, the disadvantage is that we are not leveraging the strongly typed ability of Rust. Using strongly typed data structures is great because you would want the program to fail when the incoming data format has changed, which they do a lot, and you want that information fast. You don't want corrupted data to pass through unhindered and pollute your machine learning applications downstream.

So what we will do is put in a strongly typed structure and deserialize based on that structure. Thus, we get the added benefit of data validation as well.

As an example, we will move to a little bit more complicated json file that is a dataset of Nobel Prizes in different domains.[1] To parse the dataset, we will create what is essentially a hierarchy of structs (Listing 4-6).

---

[1]Download prizes data-set http://api.nobelprize.org/v1/prize.json.

Listing 4-6. chapter4/working_with_data/data_formats/src/
jsonreading.rs

```rust
#[derive(Debug, Serialize, Deserialize)]
struct Prizes {
  prizes: Vec<Prize>,
}

#[derive(Debug, Serialize, Deserialize)]
#[allow(non_snake_case)]
struct Prize {
  category: String,
  #[serde(default)]
  overallMotivation: Option<String>,
  laureates: Vec<NobelLaureate>,
  #[serde(deserialize_with = "de_u16_from_str")]
  year: u16,
}

#[derive(Debug, Serialize, Deserialize)]
struct NobelLaureate {
  share: String,
  #[serde(default)]
  motivation: Option<String>,
  surname: String,
  #[serde(deserialize_with = "de_u16_from_str")]
  id: u16,
  firstname: String,
}
```

So the top-level field is prizes, which is a list of Prizes, which has categories, overallMotivation and year as single attributes and laureates as a list of laureates; laureates have share, motivation, surname, id, and firstname as attributes. Motivation and overallMotivation may not be present. Hence we can keep them as optional by telling serde to give default values, which in this case will be None. Since year and id are numbers, we will try to modify them to u16 and write a simple helper function de_u16_from_str to do the transformation from string to u16. By default the serde deserializer only supports conversion to string (Listing 4-7).

Listing 4-7. chapter4/working_with_data/data_formats/src/jsonreading.rs

```
fn de_u16_from_str<'a, D>(deserializer: D)
        -> Result<u16, D::Error>
        where D: Deserializer<'a>
{
  let s = String::deserialize(deserializer)?;
  u16::from_str(&s).map_err(de::Error::custom)
}
```

In Listing 4-7, deserializer is something that implements Serde's deserializer trait, and any reference to this object in memory would live for the 'a lifetime. Once the specific string object is obtained from the deserializer, we parse it to get the u16 object using u16::from_str function and then return it. If there is an error, then de::Error::custom error is raised.

Now that we are done defining the structure, serde should be able to effortlessly read the file into the structure (Listing 4-8).

Listing 4-8. chapter4/working_with_data/data_formats/src/
jsonreading.rs

```
use std::fs::File;

pub fn run() -> Result<(), Box<dyn Error>> {
  // previous code ...
  println!("from prizes json file");
  let file = File::open("data/prize.json")
    .expect("file should open read only");
  let prizes_data: Prizes = serde_json::from_reader(file)
    .expect("file should be proper JSON");

  // inspect the output
  let prizes_0 = &prizes_data.prizes[0];
  println!("category: {:?}", prizes_0.category);
}
```

The output from Listing 4-8 run should be similar to that in Listing 4-9.

Listing 4-9. chapter4/working_with_data/data_formats/src/
jsonreading.rs output

```
$ cargo run json
    Finished dev [unoptimized + debuginfo] target(s) in 0.02s
    Running `target/debug/data_formats json`
String("London")
from prizes json file
category: "physics"
```

If there are no errors, we are able to parse the json file without any issues and the schema struct that has been defined is working fine.

4.2 XML

The other popular data format is the XML data format. XML format is also considered to be one of the open standards for communication between apps and devices. Hence a lot of data is exposed in the XML format as well. The working and logic of parsing the XML file works the same way as shown in the JSON file format. We define the data in a hierarchy of structs and then try to deserialize the data using serde. In this case, to parse an XML file, we will use `serde-xml-rs` crate.

The code for parsing the XML files would be in the same binary package `data_formats`, which we have created in the previous JSON schema. This should enable us to realize that both XML and JSON, by being in inherently hierarchical data formats, will have similar code structures when writing the Rust parsing code. The only difference would be in the underlying dependencies. The rest of the code in this chapter would also embody similar motivations. Hence we will update the Cargo.toml file in the same `data_formats` binary package (Listing 4-10). Note that in addition to the crates seen before we have the serde-xml-rs crate added for parsing the XML files.

Listing 4-10. chapter4/working_with_data/data_formats/Cargo.toml

```
[dependencies]
serde = "1.0.90"
serde_derive = "1.0.90"
serde_json = "1.0"
serde-xml-rs = "0.3.1" // adding xml dependencies
```

We will create a Rust model of the `sample_2.xml` file. This file can be found in `chapter4/data/sample_2.xml`. The file structure looks something like that in Listing 4-11.

Listing 4-11. Possible file structure

```
project
    libraries
        library
    module
        files
            file
        libraries
            library
    module
        files
            file
```

Open the file and notice the above structure. So we will go ahead and create structs that capture the above structure in the xml file (Listing 4-12).

Listing 4-12. chapter4/working_with_data/data_formats/src/xmlreading.rs

```rust
use serde_xml_rs;
use serde_xml_rs::from_reader;
use serde_xml_rs::Deserializer;

#[derive(Deserialize, Debug)]
struct Project {
  name: String,
  libraries: Vec<Libraries>,
  module: Vec<Module>,
}

#[derive(Deserialize, Debug)]
struct Module {
  files: Vec<Files>,
```

```
  #[serde(default)]
  libraries: Vec<Libraries>,
}

#[derive(Deserialize, Debug)]
struct Files {
  file: Vec<FileName>,
}

#[derive(Deserialize, Debug)]
struct FileName {
  name: String,
  #[serde(rename = "type")]
  lang: String,
  #[serde(rename = "$value")]
  body: String,
}

#[derive(Deserialize, Debug)]
struct Libraries {
  library: Vec<Library>,
}

#[derive(Deserialize, Debug)]
struct Library {
  #[serde(rename = "groupId")]
  group_id: String,
  #[serde(rename = "artifactId")]
  artifact_id: String,
  version: String,
}
```

On a small technicality, in the struct FileName we cannot use the attribute name as type and need to rename from lang to type because type is a keyword in Rust. The other renamings are done for aesthetic reasons. Also #[serde(rename = "$value")] helps in getting the value of the specific field. All the other constructs should be self-explanatory as this is similar to how deserialization happens in csv and json.

We should now be able to pass the file to the deserializer, and that should create our `project` variable (Listing 4-13).

Listing 4-13. chapter4/working_with_data/data_formats/src/ xmlreading.rs

```
pub fn run() -> Result<(), Box<dyn Error>> {
  let file = File::open("data/sample_2.xml").unwrap();
  let project: Project = from_reader(file).unwrap();
  println!("{:#?}", project.libraries[0].library[0]);
  Ok(())
}
```

Similar to what we have seen in the JSON section, we will need to add the module in `main.rs` file and also add an argument so that we are able to execute the `run` function in the `xmlreading` module. Hence, we will add the following lines in the `main` method (Listing 4-14).

Listing 4-14. chapter4/working_with_data/data_formats/src/main.rs

```
// previous imports
mod jsonreading;
mod xmlreading;

fn main() {
  // previous code ...
```

```
let res = match model {
  None => {println!("nothing", ); Ok(())},
  Some("json") => jsonreading::run(),
  Some("xml") => xmlreading::run(),
  Some(_) => {println!(
    "only json and xml allowed right now", ); Ok(())},
};

// remaining part of the code ...
}
```

Notice the advantage of the strongly typed constructs in Listing 4-12. These is essentially free data validation: write once and all our data woes are gone. Also, we are able to effectively navigate through the data as if they are code. For example, something like what follows becomes possible.[2]

```
pub fn run() -> Result<(), Box<dyn Error>> {
  // previous code ...

  println!("{:#?}", project.libraries[0].library[0]);
}
```

The output of running the xmlreading::run function and seeing the output of the above print statement, we should get the output shown in Listing 4-15.

Listing 4-15. xmlreading::run output

```
$ cargo run xml
    Finished dev [unoptimized + debuginfo] target(s) in 0.03s
     Running `target/debug/data_formats xml`
Library {
    group_id: "org.example",
```

---

[2]The claim is not totally true for the given example. Can you figure out why?

```
    artifact_id: "<name>",
    version: "0.1",
}
```

In the code that is shared with the book, the csv code is also kept for completeness. The code can be downloaded from the website of the book at apress.com.

4.3 Scraping

Until now we have seen how to read csv, json, and xml file formats. Although these data formats are quite popular to store and share structured data, it might so happen that the data is not present in a structured format at all, and we might need to gather the data from different sources and collate it all together. One of the ways that can be done is through web scraping. Web scraping is about sifting through publicly available data on the internet and passing the data to a downward process. To do web scraping through Rust, we will need two crates that we will list as the dependencies, namely reqwest and scraper.

To explore the required code to perform web scraping, we will now create another package named scraping (Listing 4-16).

Listing 4-16. scraping package creation

```
$ cd chapter4 && cargo new scraping -bin
```

We can now update the Cargo.toml file with the reqwest and scraper dependencies (Listing 4-17).

Listing 4-17. chapter4/working_with_data/scraping/Cargo.toml

```
[package]
name = "scraping"
version = "0.1.0"
edition = "2018"
```

```
[dependencies]
reqwest = "0.9.15"
scraper = "0.10.0"
```

The package `reqwest` is a convenient higher-level web client, which we will use to access the specific web pages, and `scraper` will be used to parse and query the html pages using CSS selectors.

As an example, we will take the `moneycontrol.com` website. This website is an Indian online business news website. The important thing to keep in mind is that this website also publishes periodic stock prices of different public companies listed on the BSE and NSE. BSE and NSE are two major stock exchanges in India.

Now consider a hypothetical scenario where we are interested in building a dataset of the timelines and the prices of a particular Indian company, namely NTPC. NTPC is a thermal power corporation owned by the government of India and is listed on the sensex. We can find the current prices of the NTPC stock on the moneycontrol website in this money control NTPC link: `https://www.moneycontrol.com/india/stockpricequote/power-generation-distribution/ntpc/NTP`.

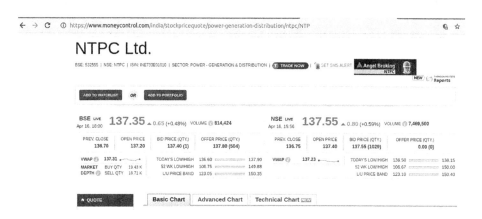

Figure 4-1. *Moneycontrol*

As can be seen in Figure 4-1, the BSE and NSE prices are listed. To get the data, we will need to get the response of the page. This can be done using the reqwest apis (Listing 4-18).

Listing 4-18. chapter4/working_with_data/scraping/src/main.rs

```
use reqwest;

fn main() -> Result<(), Box<std::error::Error>> {
  let mut resp = reqwest::get(
    "https://www.moneycontrol.com/india/stockpricequote/power-
    generation-distribution/ntpc/NTP")?;
  assert!(resp.status().is_success());

  // remaining code ...
}
```

This part of the code will store the response in the resp variable. Once done, we will need to parse the html, select the specific target data, and then collect it in a variable. This is shown in Listing 4-19.

Listing 4-19. chapter4/working_with_data/scraping/src/main.rs

```
use scraper::{Selector, Html};

fn main() -> Result<(), Box<std::error::Error>> {
  // previous code ...

  let body = resp.text().unwrap();
  let fragment = Html::parse_document(&body);
  let stories = Selector::parse("#Bse_Prc_tick > strong:nth-
  child(1)").unwrap();
```

```
for price in fragment.select(&stories) {
    let price_txt = price.text().collect::<Vec<_>>();
    // rest of the code ...

    // remaining code ...
}
```

The string that goes into the `Selector::parse` method is the css selector. We can get the appropriate selector using the chrome tools. Right-click on the specific item, in this case the BSE price, and go to `Inspect`. This should open up the chrome developer tools. Once the tool is opened, copy the selector. Take a look at Figure 4-2.

Once you copy the selector, you should get the selector string, which is the identifier for the price in the website. The app is almost complete. We can now use the time module in standard library to get a snapshot of the time when the price was captured, and this can then be printed out to stdout. We can then probably run this app in a cronjob or other scheduling system. In that way, we get a time series of the time and the price of the stock (Listing 4-20).

Figure 4-2. *Chrome Selector*

Listing 4-20. chapter4/working_with_data/scraping/src/main.src

```
use std::time::{SystemTime, UNIX_EPOCH};

fn main() -> Result<(), Box<std::error::Error>> {
  let start = SystemTime::now();

  let since_the_epoch = start.duration_since(UNIX_EPOCH)
    .expect("Time went backwards");

  // scraping code ...

  for price in fragment.select(&stories) {
    let price_txt = price.text().collect::<Vec<_>>();
      if price_txt.len() == 1 {
        println!("{:?}", (since_the_epoch, price_txt[0]));
      }
  }

  Ok(())

}
```

The output of Listing 4-20 should be something like that in Listing 4-21.

Listing 4-21. scraping output

```
$ cargo run
    Finished dev [unoptimized + debuginfo] target(s) in 0.60s
    Running `target/debug/scraping`
(1566103337.390445s, "1.05")
```

4.4 SQL

A lot of data is present in SQL databases, and hence to run machine learning algorithms on those data, we will need a way to talk to those databases and aggregate data from them. In this section we will take a look at the postgres database and try to load, access, and run an SQL query on our postgres database. For this we will use the postgres Rust crate, which as the name suggests, is the Rust native crate for `postgres`. Apart from writing native SQL queries, one other popular method of querying databases is using ORMs. Rust also has a popular ORM crate named `diesel,` but we will not look at using ORMs. This is because ORMs generally are best suited for web applications where the types of queries that developers do are fairly consistent and predictable. However, in machine learning applications, we might need to write complex queries in which case, writing queries using ORMs might become too complicated. Hence we will stick to writing native queries in SQL.

The code for this SQL section would also be in a separate package named `databases` (Listing 4-22).

Listing 4-22. Create binary package databases

```
$ cd chapter4 && cargo new databases --bin
```

Now before writing any code, we will need a postgres database. There are different ways to create the database. We could install postgres and then access the db using a client. But probably a simpler way would be to use the postgres docker. Docker is a containerization platform to package the software along with the underlying dependencies in a docker container so that it runs seamlessly in any environment. To be able to use docker, we would need to have docker installed. You can have a look at the ways of installing docker using this link: `https://docs.docker.com/v17.12/ install/#supported-platforms`. Once docker is installed, the command in Listing 4-23 can be executed to have a running postgres container. Execute the command in a separate terminal.

Listing 4-23. docker postgres

```
sudo docker run \
  --name rust-postgres \
  -e POSTGRES_PASSWORD=postgres -e POSTGRES_USER=postgres \
  -p 5432:5432 -d \
  postgres
```

The above command will run a postgres container with the name rust-postgres with user postgres and password postgres. Postgres runs on port 5432 and hence that will be mapped to the 5432 port on the host computer. The -d command means to run this container in daemon or, in other words, as a background process.

Now we can start updating the databases package. As an example, we will input the data from this url: https://www.yr.no/place/India/Karnataka/Bangalore/statistics.html. So we will create a vector of tuples with the data in a different module, postgres_db.rs (Listing 4-24).

Listing 4-24. chapter4/databases/src/postgres_db.rs

```
pub fn run() -> Result<(), Box<Error>> {
  let weathers = vec![
    ("January", 21.3, 27.3, 15.1),
    ("February", 23.6, 30.1, 17.0),
    ("March", 26.1, 32.7, 19.5),
    ("April", 28.0, 34.2, 21.8)
  ];
  // remaining part of the code ...
}
```

To feed the data, we will first need to have the postgres crate in our dependencies (Listing 4-25).

Listing 4-25. chapter4/working_with_data/databases/Cargo.toml

```
[package]
name = "databases"
version = "0.1.0"
edition = "2018"

[dependencies]
postgres = "0.15.2"
```

And then we can create the connection to postgres using the connection string, which needs to be in the format "postgresql:// user:password@uri:port/databasename". After establishing the connection, we will create the table (Listing 4-26).

Listing 4-26. chapter4/databases/src/postgres_db.rs

```
use postgres;
use postgres::{Connection, TlsMode}; // all the dependencies

pub fn run() -> Result<(), Box<Error>> {
  let conn = Connection::connect(
    "postgresql://postgres:postgres@localhost:5432/postgres",
    TlsMode::None)?; // create the connection

  conn.execute("CREATE TABLE IF NOT EXISTS weather (
    id              SERIAL PRIMARY KEY,
    month           VARCHAR NOT NULL,
    normal          DOUBLE PRECISION NOT NULL,
    warmest         DOUBLE PRECISION NOT NULL,
    coldest         DOUBLE PRECISION NOT NULL
  )", &[])?; // create the table

  // remaining code ...
}
```

The next part of the code (Listing 4-27) will be to insert the data from the vector to the table.

Listing 4-27. chapter4/databases/src/postgres_db.rs

```
pub fn run() -> Result<(), Box<Error>> {
  // previous code ...

  for weather in &weathers {
    conn.execute("INSERT INTO weather \
      (month, normal, warmest, coldest) \
      VALUES ($1, $2, $3, $4)",
      &[&weather.0, &weather.1, &weather.2, &weather.3])?;
  }

  // remaining code ...
}
```

In this example, too, we are clubbing sections such as SQL and NoSQL sections so that similar code functions can be compared with each other similar to what we have seen in JSON and XML sections. Hence similar to the previous sections, we will need the main method where we will expose the postgres_db module and execute the run function (Listing 4-28).

Listing 4-28. chapter4/databases/src/main.rs

```
use std::vec::Vec;
use std::process::exit;
use std::env::args;

mod postgres_db;

fn main() {
    let args: Vec<String> = args().collect();
    let model = if args.len() < 2 {
```

```
            None
    } else {
        Some(args[1].as_str())
    };
    let res = match model {
        None => {println!("nothing", ); Ok(())},
        Some("postgres") => postgres_db::run(),
        Some(_) =>
            println!("Only postgres allowed for now.", );
            ok(())} ,
    };
    exit(match res {
        Ok(_) => 0,
        Err(e) => {
            println!("{}", e);
            1
        }
    })
}
```

We should now be able to create the table by executing the relevant table creation SQL query by running the command `cargo run postgres`.

Retrieving the data from the database can now be done. Just executing the code written until now should have the data in the database. In a separate terminal, open the database and run a select query (Listing 4-29).

Listing 4-29. postgres queries

```
$ sudo docker exec -it rust-postgres psql --username postgres
psql (11.2 (Debian 11.2-1.pgdg90+1))
Type "help" for help.
```

```
postgres=# \dt;
          List of relations
 Schema |   Name   | Type  |   Owner
--------+----------+-------+----------
 public | weather  | table | postgres
(1 row)

postgres=# select * from weather;
 id |   month   | normal | warmest | coldest
----+-----------+--------+---------+---------
  1 | January   |   21.3 |    27.3 |    15.1
  2 | February  |   23.6 |    30.1 |      17
  .. and the remaining data
(12 rows)
```

To retrieve the values from the database, we can create a struct that has the data structure (Listing 4-30).

Listing 4-30. chapter4/databases/src/postgres_db.rs

```rust
#[derive(Debug)]
struct Weather {
  id: i32, month: String,
  normal: f64, warmest: f64,
  coldest: f64
}
```

We can now use this struct to hold the data that we retrieve using select query from the database and then use it for various purposes (Listing 4-31).

Listing 4-31. chapter4/databases/src/postgres_db.rs

```rust
pub fn run() -> Result<(), Box<Error>> {
  // previous code ...

  for row in &conn.query(
        "SELECT id, month, normal, warmest, coldest \
        FROM weather", &[])? {
    let weather = Weather {
      id: row.get(0),
      month: row.get(1),
      normal: row.get(2),
      warmest: row.get(3),
      coldest: row.get(4)
    };
    println!("{:?}", weather);
  }

  // remaining code ...
}
```

We should see all the values printed in the console when we do a `cargo` run on the package root.[3]

To complete this section on SQL, we will perform an average of the warmest column to find the warmest of the temperatures in Bangalore. The main advantage of SQL in datascience is that it allows us to not pull the data to the code but take the code to the data. This can be done by writing the business logic as SQL queries and sending the queries to the database to be executed there (Listing 4-32).

---

[3]Just to remind readers, package root is the directory where the Cargo.toml file resides.

Listing 4-32. chapter4/databases/src/postgres_db.rs

```
pub fn run() -> Result<(), Box<Error>> {
 // previous code ...

  for row in &conn.query(
        "SELECT AVG(warmest) FROM weather;", &[])? {
    let x: f64 = row.get(0);
    println!("{:?}", x);   // output 31.075 for the data input here.
  }
}
```

4.5 NoSQL

Although SQL databases are quite popular, the main drawback of SQL is that they are difficult to scale. Queries rely on the indexes to understand the relationships between different tables. Hence every time tables are updated, everything needs to be recomputed again. This puts a real bottleneck on the amount of updating you can do on a database. If the target is a fast-growing data sink, an SQL database might not be able to work. As a solution for this, NoSQL came into the picture. Although NoSQL databases are of different types, in this section we will take a look at Neo4J, which comes under the class of graph databases.

In Graph databases, nodes and edges are created, which result in finding interesting relationships between data. These nodes and edges are defined in namespaces. To understand how that works, we will take the movie-lens dataset[4] and create relationships from them. Download the ml-latest-small.zip file and unzip it in the directory. For the code in this section, we will need to expose the files through http and we can do it using a handy python command. Open a terminal, cd to the

---

[4]https://grouplens.org/datasets/movielens/.

ml-latest-small directory, and run the command shown in Listing 4-33. This is so that neo4j, which will be started in a docker comtainer, is able to pick up the files. As you hopefullyunderstand, this command relies on python3 being installed in the computer. Another simple means of creating a simple http server is using an npm http-server[5] as well. Please note to expose it on port 8000. The command should be run inside the folder where the movie-lens directory is.

Listing 4-33. Bash

```bash
$ python3 -m http.server
Serving HTTP on 0.0.0.0 port 8000 (http://0.0.0.0:8000/) ...
```

To start the neo4j database, we run the below docker command in a terminal (Listing 4-34). This should start the latest neo4j container.

Listing 4-34. Bash

```bash
sudo docker run \
  --name rust-neo4j \
  --rm --env=NEO4J_AUTH=none \
  --publish=7474:7474 \
  --publish=7687:7687 \
  --volume=$HOME/neo4j/data:/data neo4j:3.5.8
```

We will now add the ability to talk to neo4j to the databases package. We will need to add a rusted-cypher crate in the dependencies (Listing 4-35).

---

[5]https://stackoverflow.com/questions/16333790/node-js-quick-file-
server-static-files-over-http.

Listing 4-35. chapter4/databases/Cargo.toml

```
[package]
name = "databases"
version = "0.1.0"
edition = "2018"

[dependencies]
postgres = "0.15.2"
rusted_cypher = "1.1.0" // adding the neo4j dependency
```

We should now be able to connect to the database in our code. Similar to what is seen in the SQL section, the connection string needs to be passed, which is of the format `http://username:password@uri:port/db/data`. In our case we have the neo4j started without authentication and hence will not need the username and password (Listing 4-36).

Listing 4-36. chapter4/databases/src/neo4j_db.rs

```
use rusted_cypher;
use rusted_cypher::{GraphClient, Statement, GraphError};
use std::iter::repeat;

fn main() -> Result<(), Box<GraphError>> {
    let graph = GraphClient::connect(
      "http://localhost:7474/db/data")?;
    // rest of the code..
```

To load the movie data, we will first need to have the namespaces (Listing 4-37).

Listing 4-37. chapter4/databases/src/neo4j_db.rs

```
fn main() -> Result<(), Box<GraphError>> {
  // previous code ...
  let mut query = graph.query();
```

```
let statement1 = Statement::new(
  "CREATE CONSTRAINT ON (m:Movie) ASSERT m.id IS UNIQUE;");
let statement2 = Statement::new(
  " CREATE CONSTRAINT ON (u:User) ASSERT u.id IS UNIQUE;"
);
let statement3 = Statement::new(
  " CREATE CONSTRAINT ON (g:Genre) ASSERT g.name IS UNIQUE;"
);

query.add_statement(statement1);
query.add_statement(statement2);
query.add_statement(statement3);

query.send()?;

// remaining code ...
}
```

This should create the namespaces. Now we should be able to pull the data from the files (Listing 4-38).

Listing 4-38. chapter4/databases/src/neo4j_db.rs

```
fn main() -> Result<(), Box<GraphError>> {
  // import movies.csv
  graph.exec(
    "USING PERIODIC COMMIT LOAD CSV WITH HEADERS \
    FROM \"http://10.0.1.43:8000/movies.csv\" AS line \
    WITH line, SPLIT(line.genres, \"|\") AS Genres \
    CREATE (m:Movie { id: TOINTEGER(line.`movieId`), title: \
    line.`title` }) \
    WITH Genres \
    UNWIND RANGE(0, SIZE(Genres)-1) as i \
```

```
  MERGE (g:Genre {name: UPPER(Genres[i])}) \
  CREATE (m)-[r:GENRE {position:i+1}]->(g);"
)?;

// import ratings.csv
graph.exec(
  " USING PERIODIC COMMIT LOAD CSV WITH HEADERS \
  FROM \"http://10.0.1.43:8000/ratings.csv\" AS line \
  WITH line \
  MATCH (m:Movie { id: TOINTEGER(line.`movieId`) }) \
  MATCH (u:User { id: TOINTEGER(line.`userId`) }) \
  CREATE (u)-[r:RATING {rating: TOFLOAT(line.`rating`)}]->(m);"
)?;

// import tags
graph.exec(
  " USING PERIODIC COMMIT LOAD CSV WITH HEADERS \
  FROM \"http://10.0.1.43:8000/tags.csv\" AS line \
  WITH line \
  MATCH (m:Movie { id: TOINTEGER(line.`movieId`) }) \
  MERGE (u:User { id: TOINTEGER(line.`userId`) }) \
  CREATE (u)-[r:TAG {tag: line.`tag`}]->(m);"
)?;

// remaining code ...
}
```

The 10.0.1.43 ip that is used to get the data is the ip address of the machine, and in your case this ip would be different. In ubuntu the ip address can be seen by running the command hostname -I. In Mac the corresponding command is ipconfig getifaddr en0. Similar commands can be retrieved for other OS's.

The code until now should have the data in the database, and we should be able to see the graph in the neo4j console. For movies, this might be something similar to what is shown in Figure 4-3.

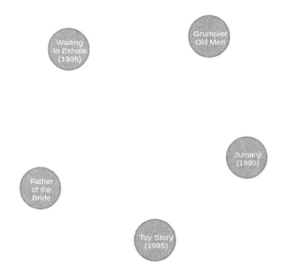

Figure 4-3. *Movie nodes*

To retrieve the data from the graph database or run queries, we can just pass the cypher queries that work in the console to the graph.exec method, and that should return the relevant results (Listing 4-39).

Listing 4-39. chapter4/databases/src/neo4j_db.rs

```
pub fn run() -> Result<(), Box<Error>> {
  // previous code ...

  let result = graph.exec(
    "MATCH (u:User {id: 119}) RETURN u.id")?;
```

```
for row in result.rows() {
  let id: u16 = row.get("u.id")?;
  println!("user id: {}", id);
}

// remaining code ...
}
```

The data can then be augmented using relational learning on the graphs [1].

4.6 Data on s3

Machine learning models, especially deep learning models, need a huge amount of data to be able to effectively train and create models that have usable levels of accuracy. Storing that amount of data might be a challenge as it we might need to have terabytes of disk space with us or invest resources in procuring and setting up that much space. Cloud storage solutions allow us to circumvent this problem. Various cloud providers such as Amazon and Azure have low-cost storage solutions that promise to store and manage our data on the internet either publicly or privately.

Amazon S3 is one of the most popular solutions in this regard. To perform S3 operations from rust, a popular crate is rusoto.[6] Rusoto is an integrated project that has a crate for each AWS service. To use S3, we will use the rusoto-s3 crate.

To explore with the code working with Amazon S3, we will create a binary package s3_files (Listing 4-40).

Listing 4-40. create binary package s3_files

```
$ cd chapter4 && cargo new s3_files --bin
```

---

[6]github source code https://github.com/rusoto/rusoto.

The folder s3_files should be created now with the files src/main.rs and Cargo.toml inside it. We can now add the S3 dependencies (rusoto_s3 and rusoto_core crates), to the toml file (Listing 4-41). We also have some additional crates such as env_logger that provides good logging configuration management, futures and futures-fs which gives good types for efficient file management. We have also seen the rustlearn crate, rand crate, csv crate and ml-utils crate before.

Listing 4-41. chapter4/s3_files/Cargo.toml

```
[package]
name = "s3_files"
version = "0.1.0"
edition = "2018"

[dependencies]
rusoto_s3 = "0.38.0"
rusoto_core = "0.38.0"
env_logger = "0.6.1"
futures = "0.1.26"
futures-fs = "0.0.5"
rand = "0.6.5"
csv = "1.0.7"
ml-utils = { path = "../../ml-utils" }
rustlearn = "0.5.0"
```

Now we should be able to write a function that takes the client, bucket, and filename and places it in S3 (Listing 4-42).

Listing 4-42. chapter4/s3_files/src/main.rs

```
use futures::{Future, Stream};
use rusoto_core;
use rusoto_core::credential::{
        AwsCredentials, DefaultCredentialsProvider};
```

```
use rusoto_core::{Region, ProvideAwsCredentials, RusotoError};
use rusoto_s3::{
    CreateBucketRequest, DeleteBucketRequest,
    DeleteObjectRequest, GetObjectRequest, ListObjectsV2Request,
    PutObjectRequest, S3Client, S3,
};

fn push_file_to_s3(
    client: &S3Client, bucket: &str,
    dest_filename: &str, local_filename: &str,
) {
  let mut f = File::open(local_filename).unwrap();
  let mut contents: Vec<u8> = Vec::new();
  match f.read_to_end(&mut contents) {
    Err(why) => panic!("Error opening file to send to S3: {}",
    why),
    Ok(_) => {
      let req = PutObjectRequest {
        bucket: bucket.to_owned(),
        key: dest_filename.to_owned(),
        body: Some(contents.into()),
        ..Default::default()
      };
      client.put_object(req).sync().expect("Couldn't PUT object");
    }
  }
}
```

The function in Listing 4-42 takes a file, denoted by local_filename, and places it in S3 where dest_filename is the key. To be able to run this function, we will need an S3 client (Listing 4-43).

Listing 4-43. chapter4/s3_files/src/main.rs

```
use std::env;
use std::fs::File;
use std::io::Read;
use std::str;
use std::vec::Vec;
use std::error::Error;

use env_logger;
use rusoto_core::{Region, ProvideAwsCredentials, RusotoError};

fn main() -> Result<(), Box<Error>> {
  let _ = env_logger::try_init();

  let region = if let Ok(endpoint) = env::var("S3_ENDPOINT") {
    let region = Region::Custom {
      // name: "us-east-1".to_owned(),
      name: "ap-south-1".to_owned(),
      endpoint: endpoint.to_owned(),
    };
    println!(
      "picked up non-standard endpoint {:?} from S3_ENDPOINT
      env. variable",
      region
    );
    region
  } else {
    // Region::UsEast1
    Region::ApSouth1
  };
  let credentials = DefaultCredentialsProvider::new()
    .unwrap().credentials()
    .wait().unwrap();
```

```
let client = S3Client::new(region.clone());
// rest of the code...
```

Now we should be able to pass the client to the push_file_to_s3 function (Listing 4-44).

Listing 4-44. chapter4/s3_files/src/main.rs

```
fn main() -> Result<(), Box<Error>> {
  // previous code ...

  let s3_bucket = format!("rust-ml-bucket");
  let filename = format!("iris.csv");
  push_file_to_s3(
    &client, &s3_bucket, &filename, "data/iris.csv");

  // remaining code ...
}
```

Running the code until now should push the file to the s3 bucket and we should see the file in s3 through the console or through the aws command line (Listing 4-45).[7]

Listing 4-45. Bash

```
$ aws s3 ls s3://rust-ml-bucket/
2019-04-21 09:28:06       3858 iris.csv
```

Now that the file is in S3, we should be able to pull the file from S3. So we create a GetObjectRequest, which the client can use to pull the data from S3. The data would be a vector of u8 characters, which we can then transform to string. Once the string is formed correctly, we can then return the string. See Listing 4-46.

---

[7]Install aws command line using the official documentation https://docs.aws.amazon.com/cli/latest/userguide/cli-chap-install.html.

Listing 4-46. chapter4/s3_files/src/main.rs

```rust
use rusoto_s3::GetObjectRequest;

fn pull_object_from_s3(client: &S3Client,
                       bucket: &str,
                       filename: &str) -> Result<String,
                                              Box<Error>> {
  let get_req = GetObjectRequest {
    bucket: bucket.to_owned(),
    key: filename.to_owned(),
    ..Default::default()
  };

  let result = client
    .get_object(get_req).sync()
    .expect("Couldn't GET object");
  println!("get object result: {:#?}", result);

  let stream = result.body.unwrap();
  let body = stream.concat2().wait().unwrap();

  Ok(str::from_utf8(&body).unwrap().to_owned())
}
```

We should now be able to pull the csv file from s3 and load into an array if Flower struct, similar to what we have seen in Chapter2 (Listing 4-47).

Listing 4-47. chapter4/s3_files/src/main.rs

```rust
use csv;
use ml_utils::datasets::Flower;

fn main() -> Result<(), Box<Error>> {
  // previous code ...
```

```
let data = pull_object_from_s3(&client, &s3_bucket,
&filename)?;

let mut rdr = csv::Reader::from_reader(data.as_bytes());
let mut data = Vec::new();
for result in rdr.deserialize() {
  let r: Flower = result?;
  data.push(r);
}

// any of the machine learning code
// that we have seen in chapter 2.
}
```

Now we should be able to put in any machine learning construct with any machine learning algorithm that we have seen in Chapter2. Apart from these, we can use other constructs in our s3 apps. Examples of creating these functions can be seen in the rusoto github repository.

4.7 Data Transformations

Writing data transformation code is quite intuitive in Rust as this would generally involve writing structs for the data constructs and then implementing methods for those constructs. We have created similar methods in Chapter 2 for different structs such as the Flower stuct and the BostonHousing struct. Moreover, we have seen how to deserialize from different data stores to structs, shown at the beginning of this chapter. Although this works even for large datasets, if the dataset is huge it may be difficult for the developer to write efficient structs for them.

Outside of the Rust ecosystem, we have the Apache Arrow ecosystem,[8] which is specifically based on the columnar data memory layout system for data, which allows data access in O(1) time. Rust developers are able to take advantage of the developments in the arrow ecosystem using the datafusion crate. Using datafusion, we can write SQL queries to perform transformations on the data. The disadvantage is that as of this writing, the datafusion library still needs a lot of functionality to be useful in a major application.

To get a basic understanding of using datafusion for parsing through datasets, we will use the titanic dataset. We can download the dataset from the kaggle website: https://www.kaggle.com/c/titanic.

We will now create the data_transformations_datafusion package using cargo (Listing 4-48).

Listing 4-48. Create binary package data_transformations_ datafusion

```
$ cd chapter4 && cargo new data_transformations_datafusion --bin
```

This would create the src/main.rs and Cargo.toml files. We will add the datafusion and arrow dependencies in Cargo.toml (Listing 4-49).

Listing 4-49. chapter4/data_transformations_datafusion/Cargo.toml

```
[package]
name = "data_transformations_datafusion"
version = "0.1.0"
edition = "2018"

[dependencies]
datafusion = "0.13.0"
arrow = "0.13.0"
```

---

[8]https://blog.cloudera.com/blog/2016/02/introducing-apache-arrow-a-fast-interoperable-in-memory-columnar-data-structure-standard/.

To be able to run the computation through arrow, we will need an execution context and a schema arc. Take a look at the code shown in Listing 4-50 to create the objects.

Listing 4-50. chapter4/working_with_data/data_transformations_ datafusion/src/main.rs

```
use arrow;
use datafusion;
use arrow::array::{
        BinaryArray, Float64Array, UInt16Array, ListArray};
use arrow::datatypes::{DataType, Field, Schema};

use datafusion::execution::context::ExecutionContext;

fn main() {
  // create local execution context
  let mut ctx = ExecutionContext::new();

  // define schema for data source (csv file)
  let schema = Arc::new(Schema::new(vec![
    Field::new("PassengerId", DataType::Int32, false),
    Field::new("Survived", DataType::Int32, false),
    Field::new("Pclass", DataType::Int32, false),
    Field::new("Name", DataType::Utf8, false),
    Field::new("Sex", DataType::Utf8, false),
    Field::new("Age", DataType::Int32, true),
    Field::new("SibSp", DataType::Int32, false),
    Field::new("Parch", DataType::Int32, false),
    Field::new("Ticket", DataType::Utf8, false),
    Field::new("Fare", DataType::Float64, false),
    Field::new("Cabin", DataType::Utf8, true),
    Field::new("Embarked", DataType::Utf8, false),
  ]));
```

```
// register csv file with the execution context
ctx.register_csv("titanic", "titanic/train.csv",
  &schema, true,
);
// rest of the code..
```

We should now be able to run queries on the data in SQL. In this case, we will find the maximum of the Fare paid by the passengers who survived. (Listing 4-51).

Listing 4-51. chapter4/working_with_data/data_transformations_ datafusion/src/main.rs

```
fn main() {
  // previous code ...

  let sql = "SELECT MAX(Fare) FROM titanic WHERE Survived = 1";

  // remaining code ...
}
```

We will then execute the query and then also pass the number of records per batch (Listing 4-52).

Listing 4-52. chapter4/working_with_data/data_transformations_ datafusion/src/main.rs

```
fn main() {
  // previous code ...

  let relation = ctx.sql(&sql, 1024 * 1024).unwrap();

  // remaining code ...
}
```

We will now need to iterate on all the batches and collect the results (Listing 4-53).

Listing 4-53. chapter4/working_with_data/data_transformations_
datafusion/src/main.rs

```rust
fn main() {
  // previous code ...

  // display the relation
  let mut results = relation.borrow_mut();

  while let Some(batch) = results1.next().unwrap() {
    println!(
      "RecordBatch has {} rows and {} columns",
      batch.num_rows(), batch.num_columns()
    );

    let name = batch
      .column(0).as_any()
      .downcast_ref::<Float64Array>().unwrap();

    for i in 0..batch.num_rows() {
      let name_value: f64 = name.value(i);

      println!("name: {}", name_value);
    }
  }
}
```

Right now, in terms of aggregation functions, there are only a limited number of functions that are implemented. More SQL features such as JOIN, ORDER BY, and LIMIT need to be implemented. Another important feature that needs work is building a dataframe api similar to the lines of apache spark.[9] Even so, we should now be able to implement a lot of common business logic using the existing features.

---

[9]https://arrow.apache.org/blog/2019/02/05/datafusion-donation/.

4.8 Working with Matrices

In this section we will explore an elementary and elegant datastructure called the matrix and its cousin, the tensor. Matrices are of immense importance in the machine learning domain as almost all of the operations and "learning" involve matrix multiplications.

Simply put, a matrix is a rectangular array of data arranged in rows and columns. Matrices support basic arithmetic operations such as addition and multiplication, and the mathematics of matrix operations are also called linear algebra. The most popular crate for linear algebra is ndarray, and we will take a look at some matrix operations in Rust using ndaray. For this we will create a binary package named `matrix_transformations` using `cargo`. See Listing 4-54.

Listing 4-54. Create binary package matrix_transformations

```
$ cd chapter4 && cargo new matrix_transformations -bin
```

We can now update the `Cargo.toml` file with the `ndarray` dependency (Listing 4-55).

Listing 4-55. chapter4/matrix_transformations/Cargo.toml

```
[package]
name = "matrix_transformations"
version = "0.1.0"
edition = "2018"

[dependencies]
ndarray-rand = "0.9.0"
ndarray = "0.12.1"
```

Let us define a 2x3 matrix, which is mathematically represented as follows.

$$A_{2x3} = \begin{bmatrix} 0 & 1 & 2 \\ 3 & 4 & 5 \end{bmatrix} \tag{1}$$

To create a matrix similar to the one above, we can create a rust vector and then create a matrix from that vector. See Listing 4-56.

Listing 4-56. chapter4/matrix_transformations/src/main.rs

```
fn main() {
  let a1 = arr2(&[[0., 1., 2.],
                  [3., 4., 5.]]);
  // remaining code ...
}
```

Printing this prints the matrix in the following fashion.
Finished dev [unoptimized + debuginfo] target(s) in 2.42s
Running `target/debug/matrix_transformations`

```
[[0.0, 1.0, 2.0],
 [3.0, 4.0, 5.0]] shape=[2, 3], strides=[3, 1], layout=C (0x1),
 const ndim=2
```

We can also use other methods such as from_vec, from_shape_vec, and from_iter according to what is convenient for us. So the below code in Listing 4-57 is equivalent to the above code in Listing 4-56.

Listing 4-57. chapter4/matrix_transformations/src/main.rs

```
fn main() {
  // previous code ...

  let a2 = Array::from_shape_vec((2, 3).strides((3, 1)),
    vec![0., 1., 2., 3., 4., 5.]).unwrap();
```

```
// remaining code ...
}
```

A common matrix transformation is taking the transpose of the matrix. Matrix transposition is a sort of mirroring where we flip the matrix along a diagonal. So the rows become the columns and the columns become the rows. Matrix transposition is needed in a lot of equations and hence is quite important. To take the transpose of a matrix in ndarray, we call the .t() method (Listing 4-58).

Listing 4-58. chapter4/matrix_transformations/src/main.rs

```
fn main() {
  // previous code ...

  let a_T = a1.t();

  // remaining code ...
}
```

Multiplying two matrices is similar to taking the dot product of them. We will multiply a1 with a_T, which are of dimensions 2x3 and 3x2, so we should have a resultant vector of 2x2 (Listing 4-59).

Listing 4-59. chapter4/matrix_transformations/src/main.rs

```
fn main() {
  // previous code ...

  let a_mm = a1.dot(&a_T);
  println!("{:?}", a_mm.shape()); // output [2, 2]

  // remaining code ...
}
```

There are a lot more operations that can be done in `ndarray`. Take a look at the exhaustive list: `https://docs.rs/ndarray/0.11/ndarray/doc/ndarray_for_numpy_users/index.html`.

4.9 Conclusion

In this chapter we introduced different ways of reading data. The different data storages we explored are the following:

- We have seen how to load data from different file formats such as json and xml.

- We have seen how to create our own structured data from open sources on the internet using web scraping.

- We have looked into pulling and storing data in SQL and NoSql data sources.

- We have also looked at BigData transformations using datafusion and matrix transformations using ndarray.

Next, we will study how to work in more real machine learning domains such as computer vision and natural language processing.

4.10 Bibliography

[1]   Maximilian Nickel et al. "A Review of Relational Machine Learning for Knowledge Graphs." In: *arXiv e-prints*, arXiv:1503.00759 (2015), arXiv:1503.00759. arXiv: `1503.00759` [`stat.ML`].

CHAPTER 5

Natural Language Processing

In the previous chapters, we took a look at different machine learning algorithms and then we worked on how to work with data for those algorithms. Those datasets were largely tabular, and the individual values were either numerical or categorical. Generally, not much processing is required in those datasets because computers are great with standardized and structured data. While trying to apply those machine learning techniques to human language though, it opens up a whole new box of challenges as language is not precise and holds different meanings in different contexts. Even the basic structure changes when we move from one language to another. Hence language needs special consideration during the creation of intelligent applications, and this is grouped under the domain of Natural Language Processing (NLP).

In this chapter we will be taking a look at different problems statements in NLP and understand the techniques that go toward solving those specific problems. First, we will take a look at sentence classification. Then we will see how to perform Named Entity Recognition on a text corpus. Finally, we will understand how to create an intent inference engine to support a good chatbot. These problems will be described using representative datasets.

© Joydeep Bhattacharjee 2020
J. Bhattacharjee, *Practical Machine Learning with Rust*,
https://doi.org/10.1007/978-1-4842-5121-8_5

5.1 Sentence Classification

To implement sentence classification in Rust, we will be using the fastText library. FastText is a library developed by Facebook for efficient learning of word representations and sentence classification. The premise is to build distributed and distributional word vectors using shallow neural networks. First, let's take a look at how to perform sentence classification on a corpus.

To work with the fastText model, we will create a package named `fasttext-model` using `cargo` similar to what we have seen in the previous chapters. This will create the `fasttext-model` directory in chapter5 folder (Listing 5-1).

Listing 5-1. Create fasttext-model package

```
$ cd chapter5 && cargo new fasttext-model --bin
$
```

To understand how to implement classification, we will need to work on a dataset. An interesting classification dataset is the spooky author dataset.[1] This dataset contains text from works of fiction in the public domain and written by three famous authors: Edgar Allan Poe, HP Lovecraft, and Mary Shelley. Download the data and unzip the files in a folder `data` inside the `fasttext-model` directory. We should now have the training file in the folder (Listing 5-2).

Listing 5-2. Peep into spooky author data

```
$ head -n2 data/train.csv
"id","text","author"
"id26305","This process, however, afforded me
no means of ascertaining the dimensions of my
dungeon; as I might make its circuit, and return
```

---

[1]https://www.kaggle.com/c/spooky-author-identification.

to the point whence I set out, without being
aware of the fact; so perfectly uniform seemed the wall.","EAP"
$

The data has three fields, so we will create a struct that reflects the data
(Listing 5-3).

Listing 5-3. chapter5/fasttext-model/src/main.rs

```
#[derive(Debug, Deserialize)]
pub struct SpookyAuthor {
  id: String, text: String, author: String
}
```

Now before moving forward, let us talk about the dependencies that we
will be using to create a simple fastText model. We have the usual suspects,
csv, serde, and serde_derive to parse the CSV file and deserialize each
record into the struct above. We have rand to shuffle between the data.
Rust packages stopwords, rust-stemmers, and vtext will be used to
normalize and tokenize the corpus. Then we will have the crate fasttext
to create the fastText classification model and itertools crate for some
helper functions. Hence, we should see the dependencies in the Cargo file
as shown in Listing 5-4.

Listing 5-4. chapter5/fasttext-model/Cargo.toml

```
[package]
name = "fasttext-model"
version = "0.1.0"
edition = "2018"

[dependencies]
csv = "1.0.7"
serde = "1"
```

```
serde_derive = "1"
rand = "0.6.5"
fasttext = "0.4.1"
stopwords = "0.1.0"
vtext = "0.1.0-alpha.1"
rust-stemmers = "1.1.0"
itertools = "0.8.0"
```

As usual, use the latest version for the different crates. Once the crates are in the dependencies, we should be able to import the relevant modules in our main.rs file (Listing 5-5).

Listing 5-5. chapter5/fasttext-model/src/main.rs

```rust
extern crate serde;
#[macro_use]
extern crate serde_derive;

use std::io;
use std::vec::Vec;
use std::error::Error;
use std::io::Write;
use std::fs::File;
use std::collections::HashSet;

use csv;
use rand::thread_rng;
use rand::seq::SliceRandom;

use fasttext::{FastText, Args, ModelName, LossName};
use stopwords::{Spark, Language, Stopwords};
use itertools::Itertools;
use vtext::tokenize::VTextTokenizer;
use rust_stemmers::{Algorithm, Stemmer};
```

We will use some constants in our code so we can define them in
Listing 5-6.

Listing 5-6. chapter5/fasttext-model/src/main.rs

```
const TRAIN_FILE: &str = "data.train";
const TEST_FILE: &str = "data.test";
const MODEL: &str = "model.bin";
```

FastText requires the training files to be in a specific format. You will
have the labels prefixed by __label__ keyword and then a space is used as
the delimiter; and once all the labels for the specific sentence or document
are done, then the rest of the sentence comes after it. For example, if l1 and
l2 are two labels for a sentence, "this is a sentence," then fastText expects it
to be in the format shown in Listing 5-7.

Listing 5-7. fastText format example

```
__label__l1 __label__l2 this is a sentence
```

Different documents are differentiated by new lines. So, we will need to
convert the present format to the fastText format. For that, of course, we will
first need to read the train.csv file to our struct. The idea is that once the
raw data is deserialized into the structs, we should be able to implement
the needed changes as methods in the struct. Similar to what was in
Chapter2, we will read the lines in the CSV file to a vector of SpookyAuthor
struct and shuffle the vector for better classification later (Listing 5-8).

Listing 5-8. chapter5/fasttext-model/src/main.rs

```
fn main() -> Result<(), Box<Error>> {
  let mut rdr = csv::Reader::from_reader(io::stdin());
  let mut data = Vec::new();
  for result in rdr.deserialize() {
    let r: SpookyAuthor = result?;
```

```
  data.push(r); // all the data is pushed to this vector.
}
data.shuffle(&mut thread_rng()); // we random shuffle the data
let test_size: f32 = 0.2; // test size is 20%
let test_size: f32 = data.len() as f32 * test_size;
let test_size = test_size.round() as usize;
let (test_data, train_data) = data.split_at(test_size);

// rest of the code...
```

Since we need to prefix the labels with the __label__ keyword, we will implement a into_labels method for SpookyAuthor (Listing 5-9).

Listing 5-9. chapter5/fasttext-model/src/main.rs

```
impl SpookyAuthor {
  // other methods..

  fn into_labels(&self) -> String {
    match self.author.as_str() {
      "EAP" => String::from("__label__EAP"),
      "HPL" => String::from("__label__HPL"),
      "MWS" => String::from("__label__MWS"),
      l => panic!("
        Not able to parse the target. \
        Some other target got passed. {:?}", l),
    }
  }

  // other methods..
}
```

Now to train a good fastText model, it is advised that you perform some text preprocessing on the raw text. Text preprocessing is primarily done to achieve text normalization, which means to convert specific areas of the text so that text is more conducive for machine learning. The techniques shown in this section are not exhaustive, nor foolproof, but are generally adopted in various contexts. Please use all text normalization techniques in your specific context to achieve optimal results.

We can convert all the text to lowercase. This is mainly applicable in Germanic languages that mainly use the Latin script for writing. A prime example of this is English. In these languages, there is a difference between lowercase and uppercase, and they are considered different during processing in a computing device. Hence it might be important to convert text to lowercase so that similar text is not considered differently during the learning process during training. But we should be careful when converting all text to lowercase as acronyms and enumerations might lose their meaning altogether when converted to lowercase. In Rust, the to_lowercase method on a string will convert all characters to lowercase (Listing 5-10).

Listing 5-10. chapter5/fasttext-model/src/main.rs

```
let lc_text = text.to_lowercase();
```

Computing devices have traditionally stored documents as a sequence of characters. Even Rust stores a random document as a Vec<u8>. But such a format does not provide any information to the machine learning algorithm. Hence a document needs to be broken down into **Linguistically Meaningful Units**. This process is called tokenization. We need to create different tokenization libraries for different languages as all languages are not equal in their organization of meaning. Generally, in English, the meaning resides in the words that are delimited by white space so it is relatively easier than other languages such as Chinese or

Japanese where orthographies might have no spaces to delimit "words" or "tokens." Even in English, care needs to be taken; for example "New Delhi" needs to be considered together to encapsulate the idea. In Rust, we can use the vtext crate if we are trying to tokenize on an English sentence (Listing 5-11).

Listing 5-11. chapter5/fasttext-model/src/main.rs

```
let tok = VTextTokenizer::new("en");
let tokens: Vec<&str> = tok.tokenize(lc_text.as_str()).
collect();
```

Sometimes in different languages, we have the word being morphed into different languages rules: for example, the tense of the sentence. For example, depending on the situation, the words "am," "are," and "is" are the same as "be." You might argue that these tense differentiations might be important to the resulting meaning but that might not be the case. The tense information might only be contributing noise to the core understanding. Stemming is an attempt to remove these ambiguities and is implemented through some heuristics that chops off the ends of words based on common prefixes and suffixes. In our case for the English language, we can use the rust-stemmers crate. This library provides rust implementations for some stemmer algorithms written in the snowball language.[2] In Listing 5-12 we will use the Stemmer struct to parse the tokens from the vtext tokenizer in Listing 5-11 and then get the stemmed tokens.

Listing 5-12. chapter5/fasttext-model/src/main.rs

```
let en_stemmer = Stemmer::create(
  Algorithm::English);
let tokens: Vec<String> = tokens.iter().map(
  |x| en_stemmer.stem(x).into_owned())
```

---

[2]https://snowballstem.org/algorithms/.

```
.collect();
let mut tokens: Vec<&str> = tokens.iter().map(
  |x| x.as_str()).collect();
```

Lastly, we will go ahead and remove the stopwords. Stopwords are commonly used words that are essentially language constructs and do not contribute in terms of meaning to the sentence. Common examples of stopwords in English are "a," "am," "the," "is," and so on. The target language that you are working on may or may not have stopwords in them. Removal of stopwords can be done using the stopwords package or you can just store a vector of the stopwords and filter out those stopwords from the tokens (Listing 5-13).

Listing 5-13. chapter5/fasttext-model/src/main.rs

```
let stops: HashSet<_> = Spark::stopwords(Language::English)
            .unwrap().iter().collect();
// notice that tokens was initialized as mutable
tokens.retain(|s| !stops.contains(s));
```

We can now join the tokens and return the whole string (Listing 5-14).

Listing 5-14. chapter5/fasttext-model/src/main.rs

```
tokens.iter().join(" ")
```

These have been the different preprocessing steps that can be implemented. This list of normalization techniques is not exhaustive and should be employed depending on the context. For example, in a financial setting, "₹200" should probably be "two hundred rupees" when normalized. Real-world text is quite complicated and might even involve different languages mixed together. In that case, text normalization would need to be more complicated and handle more corner cases for an effective ML model. The above steps, when put together in into_tokens methods, can be seen in Listing 5-15.

Listing 5-15. chapter5/fasttext-model/src/main.rs

```
impl SpookyAuthor {
  pub fn into_tokens(&self) -> String {
    // convert all to lowercase
    let lc_text = self.text.to_lowercase();

    // tokenise the words
    let tok = VTextTokenizer::new("en");
    let tokens: Vec<&str> = tok.tokenize(
      lc_text.as_str()).collect();

    // stem the words
    let en_stemmer = Stemmer::create(Algorithm::English);
    let tokens: Vec<String> = tokens.iter().map(
      |x| en_stemmer.stem(x).into_owned()).collect();
    let mut tokens: Vec<&str> = tokens.iter().map(
      |x| x.as_str()).collect();

    // remove the stopwords
    let stops: HashSet<_> = Spark::stopwords(Language::English)
        .unwrap().iter().collect();
    tokens.retain(|s| !stops.contains(s));

    // join the tokens and return
    tokens.iter().join(" ")
  }
  // remaining methods ...
```

Once done, we should be able to join the labels and the string to achieve the fastText format. We can then take the joined string and write to a file for the training data (Listing 5-16).

Listing 5-16. chapter5/fasttext-model/src/main.rs

```rust
fn push_training_data_to_file(train_data: &[SpookyAuthor],
    filename: &str) -> Result<(), Box<Error>> {
  let mut f = File::create(filename)?;
  for item in train_data {
    writeln!(f, "{} {}", item.into_labels(), item.into_
    tokens())?;
  }
  Ok(())
}
```

Consider that in Listing 5-8, we had split the full data into `train_data` and `test_data`. We can now pass the `train_data` to this function to create the training file (Listing 5-17).

Listing 5-17. chapter5/fasttext-model/src/main.rs

```rust
push_training_data_to_file(train_data.to_owned(),
  TRAIN_FILE)?;
```

Similar to the training file, we can create the test file as well. However, in the case of the test file, we should check that we are not adding the labels as well. Test files should only contain the tokens (Listing 5-18).

Listing 5-18. chapter5/fasttext-model/src/main.rs

```rust
fn push_test_data_to_file(test_data: &[SpookyAuthor],
    filename: &str) -> Result<(), Box<Error>> {
  let mut f = File::create(filename)?;
  for item in test_data {
    writeln!(f, "{}", item.into_tokens())?;
  }
  Ok(())
}
```

```
fn main() {
  // data loading code ...

  push_test_data_to_file(test_data.to_owned(),
    TEST_FILE)?;

  // remaining code ...
}
```

Now that all the preprocessing and data organization are done, we can go ahead with the training process. For that we will create an `Args` context and pass the context to the fastText struct for training. The args context will have the input file; since this is a classification problem, we will need to define the type of model as supervised and the loss function needs to be softmax. Softmax should be used because for multi-label classification. Using softmax is equivalent to predicting the distribution of labels, as softmax essentially converts them to probabilities. (Listing 5-19).

Listing 5-19. chapter5/fasttext-model/src/main.rs

```
fn main() -> Result<(), Box<Error>> {
  // previous code ...

  let mut args = Args::new();
  args.set_input(TRAIN_FILE);
  args.set_model(ModelName::SUP);
  args.set_loss(LossName::SOFTMAX);
  let mut ft_model = FastText::new();
  ft_model.train(&args).unwrap();

  // remaining code ...
}
```

So once the model is created, we can check the accuracy and save the model. This is done by creating vector `preds` by running the `predict` method of the fastText model on the items of `test_data`. Another variable test_labels is created by running the into_labels method on the items of test_data. These two vectors are then compared one by one in a for loop, and if they are the same then a variable correct_hits is incremented. For all values in preds, the variable hits is incremented to get a count of total values. The ratio of these two values is then computed to get the accuracy parameter (Listing 5-20).

Listing 5-20. chapter5/fasttext-model/src/main.rs

```rust
fn main() -> Result<(), Box<Error>> {
  // previous code ...

  let preds = test_data.iter().map(
    |x| ft_model.predict(x.text.as_str(), 1, 0.0));
  let test_labels = test_data.iter().map(
    |x| x.into_labels());
  let mut hits = 0;
  let mut correct_hits = 0;
  let preds_clone = preds.clone();
  for (predicted, actual) in preds.zip(test_labels) {
    let predicted = predicted?;

    // only taking the first value.
    let predicted = &predicted[0];
    if predicted.clone().label == actual {
      correct_hits += 1;
    }
    hits += 1;
  }
  assert_eq!(hits, preds_clone.len());
  println!("accuracy={} ({}/{} correct)",
    correct_hits as f32 / hits as f32,
```

```
  correct_hits, preds_clone.len());
ft_model.save_model(MODEL)?;

Ok(())
}
```

We should now be able to run the model and train the fastText classifier (Listing 5-21).

Listing 5-21. fasttext-model training

```
$ cd chapter5/fasttext-model
$ cargo run < data/train.csv
   Compiling fasttext-model v0.1.0 (chapter5/fasttext-model)
    Finished dev [unoptimized + debuginfo] target(s) in 1.42s
     Running `target/debug/fasttext-model`
Read 0M words
Number of words:  5560
Number of labels: 3
Progress: 100.0% words/sec/thread:  134810 lr:  0.000000
loss:   1.079427 ETA:    0h 0m
accuracy=0.48263535 (1890/3916 correct)
$
```

And we should be able to see the model file saved in the directory (Listing 5-22).

Listing 5-22. Trained fastText model

```
$ ls -ltr model.bin
$
-rw-r--r--  1 joydeepbhattacharjee  staff  802313179 Aug 20
09:37 model.bin
```

As this model is built in the `fastText` code library, the model.bin file generated in Listing 5-22 conforms to the general `fastText` specification and can be used by a different `fastText` application as well. For example, if we load the model using the original `fastText` application, it works just fine. See Listing 5-23.

Listing 5-23. fastText predict using the official binary

```
$ ./fasttext predict ../model.bin -
I love ghosts.
__label__EAP
^C
$
```

To compile and create the above fasttext binary, follow the instructions in the fasttext webpage: `https://github.com/facebookresearch/ fastText#building-fasttext-using-make-preferred.`

5.2 Named Entity Recognition

One of the core areas of implementation of natural language processing is Named Entity Recognition (NER), where entities that are present in the text are classified into predefined categories. These categories are context and problem dependent. For example, a travel organization may be interested in the cities and dates. You could argue that this can be done using regex, but if you are a growing company, it would be very difficult to scale up such an operation. Also going through the NER route adds a wealth of semantic knowledge to the content and helps to understand the subject of any given text.

One of the popular algorithms that is used for NER tasks are Conditional Random Fields (CRFs). CRFs are essentially classifiers that use contextual information from previous labels, thus increasing the amount of information that the label has to make a good prediction.

To perform CRF effectively, the input text would need to be chunked correctly. Text chunking divides each sentence into syntactically correlated parts of words. For example, the sentence, "Rust is great for machine learning" can be divided as follows:

[NP Rust][VP is][NP great][PP for][NP machine learning]

In this example, NP stands for a noun phrase, VP for a verb phrase, and PP for a prepositional phrase. This task is formalized as a sequential labeling task in which a sequence of tokens in a text is assigned with a sequence of labels. In order to present a chunk, the IOB2 notation is used. The beginning of a chunk is given by a B-label, inside of the chunk is given by an I-label, and others are defined as O. Chunking is mostly a manual task and there are some popular annotation tools that make it a little simpler for the user.

- GATE - General Architecture and Text Engineering is 15+ years old, free and open source.

- Anafora - It is a free and open source, web-based raw text annotation tool.

- brat - Brat rapid annotation tool is an online environment for collaborative text annotation.

- tagtog – It is proprietary tool costing money.

- prodigy - It is an annotation tool powered by active learning and costs money.

- LightTag - LightTag is a hosted and managed text annotation tool for a team and costs money.[3]

---

[3]https://github.com/keon/awesome-nlp#annotation-tools

We will, on the other hand, use a dataset that has been annotated for us. This is can be downloaded from the kaggle website with this link: `https://www.kaggle.com/abhinavwalia95/entity-annotated-corpus`. Download the ner.csv file and keep it in the data folder. We can see that the data has a lot of fields and this is a good guideline for the type of fields that should be there in an NER dataset. Let's take a look at the fields.

- lemma - Lemma of a token in sentence

- next-lemma Lemma of next token in sentence

- next-next-lemma Lemma of token at +2nd position to the current token in sentence

- next-next-pos POS tag of token at +2nd position to the current token in sentence

- next-next-shape Shape of token at +2nd position to the current token in sentence

- next-next-word Token at +2nd position to the current token in sentence

- next-pos POS tag of the next(+1 position) token

- next-shape Shape of the next(+1 position) token

- next-word Next(+1 position) token

- pos POS tag of current token

- prev-iob IOB annotation of previous token

- prev-lemma Lemma of previous token

- prev-pos POS tag for previous token

- prev-prev-iob IOB annotation of token at -2nd position to the current token in sentence

- prev-prev-lemma Lemma of token at -2nd position to the current token in sentence

- prev-prev-pos POS tag of token at -2nd position to the current token in sentence

- prev-prev-shape Shape of token at -2nd position to the current token in sentence

- prev-prev-word Token at -2nd position to the current token in sentence

- prev-shape Shape of previous (-1 position to current token) token

- prev-word Previous word (-1 position to current token)

- sentence_idx Sentence Index (Tokens having same index belongs to same sentence)

- shape Shape of the token in sentence

- word often termed as Token

- tag IOB annotation of current token

For our example, though, we will only be using the lemma and next-lemma to have an understanding of how to create an NER model using crf-suite. In a production environment, though, you should use all the above features and all the other features that you can integrate.

Now to create the code for the NER, we will create a binary package crfsuite-model using cargo (Listing 5-24).

Listing 5-24. create crfsuite-model package

```
$ cargo new crfsuite-model --bin
$
```

This should create the src/main.rs and Cargo.toml files in the directory crfsuite-model (Listing 5-25).

Listing 5-25. chapter5/crfsuite-model/src/main.rs

```
#[derive(Debug, Deserialize, Clone)]
pub struct NER {
  lemma: String,
  #[serde(rename = "next-lemma")]
  next_lemma: String,
  word: String,
  tag: String
}
```

Notice that in Listing 5-25, we have used the rename feature in serde to identify next_lemma in the NER struct to be the next-lemma column in the csv.

Before we go ahead, let's talk about the dependencies that we will need for creating the model. The crate csv is for parsing the csv files where serde and serde-derive will help as well. The crate rand is used for generating random variables and crfsuite is the main crate that can be used to have CRF capabilities in the code. See Listing 5-26.

Listing 5-26. chapter5/crfsuite-model/Cargo.toml

```
[dependencies]
csv = "1.0.7"
serde = "1"
serde_derive = "1"
rand = "0.6.5"
crfsuite = "0.2.6"
```

Once these are in the toml file, we should be able to import the relevant modules in our main module (Listing 5-27).

Listing 5-27. chapter5/crfsuite-model/src/main.rs

```
extern crate serde;
#[macro_use]
extern crate serde_derive;
```

205

```
use std::io;
use std::vec::Vec;
use std::error::Error;

use csv;
use rand;
use rand::thread_rng;
use rand::seq::SliceRandom;

# below are the new ones and useful for NER
use crfsuite::{Model, Attribute, CrfError};
use crfsuite::{Trainer, Algorithm, GraphicalModel};
```

Similar to what we have seen before, we can now create a function that reads the data from the standard input and stores it in a vector of the data struct that we have, in this case that being Vec<NER>. See Listing 5-28.

Listing 5-28. chapter5/crfsuite-model/src/main.rs

```
fn get_data() -> Result<Vec<NER>, Box<Error>> {
  let mut rdr = csv::Reader::from_reader(io::stdin());
  let mut data = Vec::new();
  for result in rdr.deserialize() {
    let r: NER = result?;
    data.push(r);
  }
  data.shuffle(&mut thread_rng());
  Ok(data)
}
```

Given that now we have the data, we should be able to split the data into train and test, so that we can check the accuracy on out-of-sample data. This is the same code that we have seen before in other modules and chapters (Listing 5-29).

Listing 5-29. chapter5/crfsuite-model/src/main.rs

```
fn split_test_train(data: &[NER],
                    test_size: f32)
                    -> (Vec<NER>, Vec<NER>) {
  let test_size: f32 = data.len() as f32 * test_size;
  let test_size = test_size.round() as usize;
  let (test_data, train_data) = data.split_at(test_size);
  (test_data.to_vec(), train_data.to_vec())
}
```

Now comes the interesting part. Given the data, we will need to extract each item in the data to a vector of Attributes, provided by the crf_suite crate so that the training module is able to train on the dataset. This Attribute will contain the token and the value of the token, which is the weightage that the token has in the sequence. In the Listing 5-30, weightage of 1.0 is given for the target word and 0.5 for the next word, but you can play with these weightage and see which gives a better result. The labels can be a Vec<String> as expected.

Listing 5-30. chapter5/crfsuite-model/src/main.rs

```
fn create_xseq_yseq(
    data: &[NER])
    -> (Vec<Vec<Attribute>>, Vec<String>) {
  let mut xseq = vec![];
  let mut yseq = vec![];
  for item in data {
    let seq = vec![Attribute::new(item.lemma.clone(), 1.0),
      Attribute::new(item.next_lemma.clone(), 0.5)];
      // higher weightage for the mainword.
    xseq.push(seq);
```

```
    yseq.push(item.tag.clone());
  }
  (xseq, yseq)
}
```

We can now create a function that will do the model prediction given x-sequence and y-sequence. In this case, after training, the model saves in a file that is determined by model_name (Listing 5-31).

Listing 5-31. chapter5/crfsuite-model/src/main.rs

```
fn crfmodel_training(xseq: Vec<Vec<Attribute>>,
                     yseq: Vec<String>,
                     model_name: &str)
                     -> Result<(), Box<CrfError>> {
  let mut trainer = Trainer::new(true); // verbose is true
  trainer.select(Algorithm::AROW, GraphicalModel::CRF1D)?;
  trainer.append(&xseq, &yseq, 0i32)?;
  trainer.train(model_name, -1i32)?; // using all instances for
                                                      training.
  Ok(())
}
```

In Listing 5-31, we used the Adaptive regularization of weight vector. The different algorithms that we can use are the following:

- Gradient descent using the L-BFGS method: It is a way of finding the local minimum of objective function, and making use of objective function values and the gradient of the objective function.

- Stochastic Gradient Descent with L2 regularization term: It combines stochastic gradient descent with the loss function having an additional parameter of the square of the weights. This should avoid the model having to overfit on the data.

- Averaged Perceptron: This is a binary classification method where averaged weights and biases of a standard perceptron algorithm are used.

- Passive Aggressive: This algorithm slowly forgets the old distribution in case the data is taken from a different sample. It is passive if a correct classification occurs and "aggressive" or there is a change in weights in case there is a missclassification.

- Adaptive Regularization of Weight Vector: This algorithm combines large margin training, confidence weighting, and the capacity to handle non-separable data. AROW performs adaptive regularization of the prediction function upon seeing each new instance, allowing it to perform especially well in the presence of label noise.

Once training is done, we can load the model from file and run predictions on it (Listing 5-32).

Listing 5-32. chapter5/crfsuite-model/src/main.rs

```rust
fn model_prediction(xtest: Vec<Vec<Attribute>>,
                    model_name: &str)
                        -> Result<Vec<String>, Box<CrfError>>{
  let model = Model::from_file(model_name)?;
  let mut tagger = model.tagger()?;
  let preds = tagger.tag(&xtest)?;
  Ok(preds)
}
```

We can use this model_prediction function and see how our model fared on the test data. To do this, we need to use the accuracy function. The accuracy function here is the same as the ones that we have seen in previous chapters. See Listing 5-33.

Listing 5-33. chapter5/crfsuite-model/src/main.rs

```
fn check_accuracy(preds: &[String], actual: &[String]) {
  let mut hits = 0;
  let mut correct_hits = 0;
  for (predicted, actual) in preds.iter().zip(actual) {
    if actual != "O" { // will not consider the other category
                       as it bloats the accuracy.
      if predicted == actual && actual != "O" {
        correct_hits += 1;
      }
      hits += 1;
    }
  }
  println!("accuracy={} ({}/{} correct)",
    correct_hits as f32 / hits as f32,
    correct_hits,
    hits);
}
```

Notice the difference: we will not consider the values where the actual is "O." That is because this label means that other and most of the labels would be others. So considering this as part of the accuracy, values would essentially bloat up our accuracy and give accuracy results better than they actually are. A more general way in which the accuracy of the CRF model is determined is by using precision and recall and the F1 score, which can also be implemented in Rust in a similar manner.

210

Now that we have all the relevant functions, we can stitch all those functions in our main function (Listing 5-34).

Listing 5-34. chapter5/crfsuite-model/src/main.rs

```
fn main() {
  let data = get_data().unwrap();
  let (test_data, train_data) = split_test_train(
      &data, 0.2);
  let (xseq_train, yseq_train) = create_xseq_yseq(
      &train_data);
  let (xseq_test, yseq_test) = create_xseq_yseq(
      &test_data);
  crfmodel_training(xseq_train,
      yseq_train,
      "rustml.crfsuite").unwrap();
  let preds = model_prediction(xseq_test,
      "rustml.crfsuite").unwrap();
  check_accuracy(&preds, &yseq_test);
}
```

Running the code in Listing 5-34 should have an output similar to that shown in Listing 5-35.

Listing 5-35. crfsuite-model output

```
$ cd chapter5/crfsuite-model
$ cargo run < data/ner.csv
    Finished dev [unoptimized + debuginfo] target(s) in 0.03s
    Running `target/debug/crfsuite-model`
Feature generation
type: CRF1d
feature.minfreq: 0.000000
```

```
feature.possible_states: 0
feature.possible_transitions: 0
0....1....2....3....4....5....6....7....8....9....10
Number of features: 3137
Seconds required: 0.013

Adaptive Regularization of Weights (AROW)
variance: 1.000000
gamma: 1.000000
max_iterations: 100
epsilon: 0.000000

***** Iteration #1 *****
Loss: 1210.000000
Feature norm: 0.526780
Seconds required for this iteration: 0.012

**** Intermediate iterations

***** Iteration #100 *****
Loss: 369.217403
Feature norm: 296.558922
Seconds required for this iteration: 0.012

Total seconds required for training: 1.226

Storing the model
Number of active features: 3116 (3137)
Number of active attributes: 2067 (2088)
Number of active labels: 17 (17)
Writing labels
Writing attributes
Writing feature references for transitions
```

```
Writing feature references for attributes
Seconds required: 0.014

accuracy=0.5551948 (171/308 correct)
$
```

And we should have the model being created in the directory (Listing 5-36).

Listing 5-36. model output file

```
$ ls -ltr rustml.crfsuite
-rw-r--r--   1 joydeepbhattacharjee  staff  168196 Aug 20 09:10
rustml.crfsuite
```

As you can see in Listing 5-35, the accuracy is quite low. That is because we have utilized only three features for the sake of simplicity. Try with more features and see if the accuracy improves or not.

Named Entity Recognitions are an important and interesting set of problems in NLP, and the crfsuite is an excellent crate to create a CRF model for NER.

In this section, we looked at how to use to create a CRF model for NER in Rust. We also looked at how to mold an incoming dataset so that it is conducive to the CRF suite model. Finally, we took a look at one of the ways in which accuracy for a CRF model can be decided.

5.3 Chatbots and Natural Language Understanding (NLU)

While chatbots is a complex topic and involves a lot of integration and engineering for a production system, one of the core areas that powers a chatbot is for the underlying system to have some level of natural language understanding. NLU is needed to determine a user's intention and to extract information from an utterance and to carry on a conversation with the user in order to execute and complete a task.

Intent Recognition   Most intents are simple discrete tasks like "Find Product," "Transfer Funds," "Book Flight," and are typically described with the verb and noun combination. These types of intents will initiate a dialogue with the user to capture more information, to fetch and update data from remote systems, and to inform the user of its progress.

The goal of intent recognition is to match a user utterance with its correctly intended task or question. Intent parsing of user utterance can be done in Rust using the crate `snips-nlu-lib`. The caveat though is that this library can only be used for the inference part, and training needs to be done in python.

Now before moving ahead, ensure that you have clang installed in your machine. For an ubuntu machine, installation of clang can be done using the command shown in Listing 5-37.

Listing 5-37. ubuntu install clang

```
$ sudo apt update
$ sudo apt install clang
# type yes and <enter> in case there is a prompt
$
```

In case you are using a mac, it should have clang installed or you can install it using Xcode.[4]

Let's talk about how to use steps to create a snips model. Since the snips model is a python model, we can create a virtual environment and install the package (Listing 5-38).

---

[4]https://developer.apple.com/xcode/.

Listing 5-38. Create python virtual environment

```
$ python3 -m venv venv
$ source venv/bin/activate
(venv) $ pip install snips-nlu
(venv) $
```

To download the model, we will need to download the English model. Other models as of this writing are German, Spanish, French, Italian, Japanese, Korean, and Portuguese (one for Brazil dialect and one for Portugal dialect).[5] Implementing a different language involves changes in a lot of repositories, which is not very simple and hence the snips team is working on making the whole process simpler.[6] If downloading the English model does not work for the remaining steps, we can also download all the resources and try again (Listing 5-39).

Listing 5-39. Download english

```
(venv) $ python -m snips_nlu download en
# or we can try
(venv) $ python -m snips-nlu download-all-languages
(venv) $
```

We can now create a dataset.yaml for our different types of utterances. See Listing 5-40.

Listing 5-40. contents of dataset.yaml

```
$ cat dataset.yaml
# turnLightOn intent
---
```

---

[5]https://snips-nlu.readthedocs.io/en/latest/languages.html.

[6]https://github.com/snipsco/snips-nlu-language-resources/issues/12#iss
  uecomment-433325114.

```
type: intent
name: turnLightOn
slots:
  - name: room
    entity: room
utterances:
  - Turn on the lights in the [room](kitchen)
  - give me some light in the [room](bathroom) please
  - Can you light up the [room](living room) ?
  - switch the [room](bedroom)'s lights on please

# turnLightOff intent

---
type: intent
name: turnLightOff
slots:
  - name: room
    entity: room
utterances:
  - Turn off the lights in the [room](entrance)
  - turn the [room](bathroom)'s light out please
  - switch off the light the [room](kitchen), will you?
  - Switch the [room](bedroom)'s lights off please

# setTemperature intent

---
type: intent
name: setTemperature
slots:
  - name: room
    entity: room
  - name: roomTemperature
    entity: snips/temperature
```

```
utterances:
  - Set the temperature to [roomTemperature](19 degrees) in the
    [room](bedroom)
  - please set the [room](living room)'s temperature to
    [roomTemperature](twenty two degrees celsius)
  - I want [roomTemperature](75 degrees fahrenheit) in the
    [room](bathroom) please
  - Can you increase the temperature to [roomTemperature]
    (22 degrees) ?

# room entity
---
type: entity
name: room
automatically_extensible: no
values:
- bedroom
- [living room, main room, lounge]
- [garden, yard, backyard]
$
```

We would now need to create our dataset.json file from the utterances yaml file (Listing 5-41).

Listing 5-41. Create dataset.json

```
(venv) $ snips-nlu generate-dataset en dataset.yaml > dataset.
json
(venv) $
```

Once the dataset.json file is created, we should be able to train a model (Listing 5-42).

Listing 5-42. snips training

```
(venv) $ snips-nlu train dataset.json snips.model -v
(venv) $
```

The output of the command in Listing 5-42 should be something like what is shown in Listing 5-43.

Listing 5-43. snips training output

```
Create and train the engine...
[INFO][21:55:51.091][snips_nlu.intent_parser.deterministic_
intent_parser]: Fitting deterministic parser...
[INFO][21:55:51.103][snips_nlu.intent_parser.deterministic_
intent_parser]: Fitted deterministic parser in 0:00:00.011588
[INFO][21:55:51.103][snips_nlu.intent_parser.probabilistic_
intent_parser]: Fitting probabilistic intent parser...
[DEBUG][21:55:51.103][snips_nlu.intent_classifier.log_reg_
classifier]: Fitting LogRegIntentClassifier...
[DEBUG][21:55:51.616][snips_nlu.intent_classifier.log_reg_
classifier]: Top 50 features weights by intent:

<training on different intents>
...
< training on transition weights and feature weights.>

[DEBUG][21:55:57.292][snips_nlu.slot_filler.crf_slot_filler]:
Fitted CRFSlotFiller in 0:00:01.482648
[DEBUG][21:55:57.293][snips_nlu.intent_parser.probabilistic_
intent_parser]: Fitted slot fillers in 0:00:05.671146
[INFO][21:55:57.294][snips_nlu.intent_parser.probabilistic_
intent_parser]: Fitted probabilistic intent parser in
0:00:06.190487
```

```
[INFO][21:55:57.294][snips_nlu.nlu_engine.nlu_engine]: Fitted
NLU engine in 0:00:07.315197
Persisting the engine...
Saved the trained engine
```

5.3.1 Building an Inference Engine

Once the training is done, we should be able to use the model in our application. As an example application, we will be creating a simple API that would take a sentence as an input and provide the parsed intents in the sentence.

As for the dependencies, we would be using the snips-nlu-lib crate for the snips model inference. serde, and serde_json and serde_derive would be used for serializing and deserializing the incoming request. For creating the web-app, we would be using the rocket web framework. The API from the rocket framework is inspired out of great web frameworks such as Rails, Flask, Bottle, and Yesod; and hence it is quite fun to create applications in Rocket. To create the inference engine and to play with these dependencies, we would now need to create a Rust project using cargo (Listing 5-44).

Listing 5-44. Create snips-model Rust project

```
$ cd chapter5
$ cargo new snips-model –bin
```

This should create the main.rs and Cargo.toml files in the directory and we should be able to update the Cargo file similar to what is shown in Listing 5-45. The crate serde-json will be used to parse the json files and serde and serde-derive will help towards that. The crates rocket and rocket-contrib will be utilised to create the web apis and snips-nlu-lib is the main crate giving us NLU capabilities.

Listing 5-45. chapter5/snips-model/Cargo.toml

```
[package]
name = "snips-model"
version = "0.1.0"
edition = "2018"

[dependencies]
snips-nlu-lib = { git = "https://github.com/snipsco/snips-nlu-
rs", branch = "master" }
rocket = "0.4.0"
rocket_contrib = "0.4.0"
serde = "1.0"
serde_json = "1.0"
serde_derive = "1.0"
```

Note that we are using the github link directly for the snips library as that works the best for now.

We can now import the relevant modules from the dependencies in our main file (Listing 5-46).

Listing 5-46. chapter5/snips-model/src/main.rs

```
#![feature(proc_macro_hygiene, decl_macro)]

#[macro_use] extern crate rocket;
#[macro_use] extern crate rocket_contrib;
#[macro_use] extern crate serde_derive;
extern crate snips_nlu_lib;

use std::sync::Mutex;

use snips_nlu_lib::SnipsNluEngine;
use rocket::{Rocket, State};
use rocket_contrib::json::Json;
```

We can now write a small function to see if our setup until now is working or not (Listing 5-47).

Listing 5-47. chapter5/snips-model/src/main.rs

```
#[get("/")]
fn hello() -> &'static str {
  "Hello, from snips model inference!"
}
```

To activate this endpoint, we will need to do a little setup and launch the app (Listing 5-48).

Listing 5-48. chapter5/snips-model/src/main.rs

```
fn rocket() -> Rocket {
    rocket::ignite()
        .mount("/", routes![hello])
}

fn main() {
    rocket().launch();
}
```

Run the application using cargo run. This should build all the dependencies, and then if everything hah been fine until now, you should run the web application and be able to see the following output shown in Listing 5-49.

Listing 5-49. Run snips model inference

```
$ cd chapter5/snips-model
$ cargo run
   Compiling snips-model v0.1.0 (...)
    Finished dev [unoptimized + debuginfo] target(s) in 1m 39s
     Running `target/debug/snips-model`
```

```
Loading the nlu engine...
 Configured for development.
    => address: localhost
    => port: 8000
    => log: normal
    => workers: 8
    => secret key: generated
    => limits: forms = 32KiB
    => keep-alive: 5s
    => tls: disabled
Mounting /:
    => GET / (hello)
```

Rocket has launched from http://localhost:8000

```
$
```

And we should be able to make a simple call to the webserver (Listing 5-50).

Listing 5-50. curl to snips nlu engine

```
$ curl localhost:8000/
Hello, from snips model inference!
$
```

Now we are moving into the main inference part. We will try to create a function that will handle the post request. Hence we will create a struct Message that will hold the incoming request data and a function infer, which will handle the inference part. So writing code for handling the incoming function will look something like what is in Listing 5-51.

Listing 5-51. chapter5/snips-model/src/main.rs

```
#[derive(Serialize, Deserialize)]
struct Message {
    contents: String
}

#[post("/infer", format = "json", data = "<message>")]
fn infer(message: Json<Message>,
        <remaining arguments>) -> OutputType {
    let query = message.0.contents; // to get the query from
    the incoming data
    // remaining code ...
}

fn rocket() -> Rocket {
    // Have Rocket manage the engine to be passed to the
    functions.
    rocket::ignite()
        .mount("/", routes![hello, infer])
}

fn main() {
    rocket().launch();
}
```

We will now need to load the engine, an object of type
SnipsNluEngine, run the inference based on the incoming query, and
return the result. Since this is dependent on the user query sentence, a
naive way might be to put all the code in the infer function. This however
is not ideal as creating the engine from the model is quite expensive and
having to do that all over again for each request is not ideal. Hence we
need a mechanism to load the model and create the engine once at the
start of the app initiation and then use the same model for each request

inference cycle. Since this model will be shared between threads in the Rocket app, we cannot just create a global variable or something similar. An elegant solution for this is by using Rust mutexes.[7] This is a highly useful, mutual-exclusion primitive implemented as a struct to be used for protecting shared data.

So we will wrap the engine as a mutex (Listing 5-52).

Listing 5-52. chapter5/snips-model/src/main.rs

```
type Engine = Mutex<SnipsNluEngine>;
```

A different function will be defined for creating the engine. This function will be called before initializing the Rocket app, and the engine object would be passed to the rocket app to be managed by it (Listing 5-53).

Listing 5-53. chapter5/snips-model/src/main.rs

```
fn init_engine() -> SnipsNluEngine {
  let engine_dir = "path to snips model/snips.model";
  println!("\nLoading the nlu engine...");
  let engine = SnipsNluEngine::from_path(engine_dir).unwrap();
  engine
}

fn rocket() -> Rocket {
  // load the snips ingerence engine.
  let engine = init_engine();

  // Have Rocket manage the engine to be passed to the
  functions.
  rocket::ignite()
    // let rocket manage the state of the engine
```

---

[7]https://doc.rust-lang.org/std/sync/struct.Mutex.html.

```
    .manage(Mutex::new(engine))
    .mount("/", routes![hello, infer])
}
```

Once our rocket app has access to the engine object, we can call it in our functions, `infer` in this case, and get the intents from the user-provided document (Listing 5-54).

Listing 5-54. chapter5/snips-model/src/main.rs

```
#[post("/infer", format = "json", data = "<message>")]
fn infer(message: Json<Message>, engine: State<Engine>) ->
String {
  let query = message.0.contents;

  // locking the mutex so that nothing else
  // changes it and we can utilise the underlying resource
  let engine = engine.lock().unwrap();

  // get the intents
  let result = engine.get_intents(query.trim()).unwrap();

  // serialize the result as a json string.
  let result_json = serde_json::to_string_pretty(
        &result).unwrap();

  result_json
}
```

Building the intent inference is done. We can now run the app using `cargo run` as in Listing 5-55. So apart from the other output that we have seen before when running the app, we should see the `infer` endpoint being loaded as well. We should also see the print statement that tells us that the model has been loaded from the `init_engine` function.

Listing 5-55. Launch nlu engine

```
$ cd chapter5/snips-model
$ cargo run
   Compiling snips-model v0.1.0 (...)
    Finished dev [unoptimized + debuginfo] target(s) in 1m 39s
     Running `target/debug/snips-model`

Loading the nlu engine...
// other output...
Mounting /:
    => GET / (hello)
    => POST /infer application/json (infer)
Rocket has launched from http://localhost:8000
$
```

If the infer endpoint is loaded correctly, we should now be able to make requests to the app and get the result (Listing 5-56).

Listing 5-56. Output of nlu engine in separate terminal

```
$ curl --header "Content-Type: application/json" \
        --request POST \
        --data '{"contents":"set the temperature to 23 degrees
          in the bedroom"}' \
        localhost:8000/infer
[
  {
    "intentName": "setTemperature",
    "confidenceScore": 1.0
  },
```

```
{
  "intentName": null,
  "confidenceScore": 0.291604
},
{
  "intentName": "turnLightOn",
  "confidenceScore": 0.08828778
},
{
  "intentName": "turnLightOff",
  "confidenceScore": 0.070788406
}
]
$
```

One of the important problem statements in NLP is chatbots and to build chatbots, we need a good intent inference engine. In this section, we took a look at how to train a snips model and use the model in a simple webapp for inferring the intents in a user-specified query.

5.4 Conclusion

This chapter introduced you to different interesting problems in NLP. The chapter started with text classification and how to build a classifier using the fasttext algorithm. It also had common text preprocessing steps that can be done in Rust. Then the chapter went on with a look at performing named entity recognition on a corpus and how to create a CRF model using the crfsuite crate to perform the NER task. Last, we looked at creating an intent inference web application using the snips library.

In the next chapter, you will learn about creating computer vision models and running image classifiers.

CHAPTER 6

Computer Vision

In the last chapter, we took at look at how machine learning applies in the domain of human languages. That covers our senses of speech and sound. Another sense where machine learning has a huge influence is our sense of vision and how machine learning is able to capture the essence of images. The overall branch is termed computer vision and there have been great strides in solving computer vision problems in Rust as we will see in this chapter.

As part of this chapter, we will look at different applications in computer vision such as the following:

- Image classification

- Using pretrained models

- Neural style transfer

- Face detection

6.1 Image Classification

Image classification is the application of machine learning to images, where the objective is that given an image and a model trained on finding objects of interest, the model would identify if the object is present in the

© Joydeep Bhattacharjee 2020
J. Bhattacharjee, *Practical Machine Learning with Rust*,
https://doi.org/10.1007/978-1-4842-5121-8_6

image or not. This is achieved by defining a set of target classes (objects to identify in images), and a model is trained to identify them using labeled example photos.

Image classification using traditional models was achieved by adding new features from pixel data, such as color histograms, textures, and shapes. The downside of this approach was that feature engineering in the case of image classification was a huge cost. Features needed to be tuned quite precisely, and hence building robust models was very challenging with low accuracy in many cases.

6.1.1 Convolutional Neural Networks (CNN)

In contrast with the traditional models, Neural networks with convolutional layers was seen to have better results for the image classification tasks. A CNN progressively extracts higher and higher-level representations of the image content. Instead of preprocessing the data to derive features like textures and shapes, a CNN takes just the raw pixels of the image as input and "learns" how to extract those features, and ultimately infer what object they constitute.

The CNN starts with an input feature map of three dimensions where the size of the first two dimensions is the length and width of the images in pixels. The size of the third dimension is the number of channels in the image and is generally 3. The CNN comprises stacks of modules, where each module performs three operations [1].

1. A convolution layer is where tiles of the input feature map are extracted and filters are applied to the features. A filter matrix is slid over the feature matrix and the sum of the element-wise multiplication is stored in the resulting matrix. During training the CNN "learns" the optimal value for the feature matrix, which enables it to extract meaningful

features. Figure 6-1 shows how this process takes place on a 4x4 input feature map with the filter layer being 3x3.

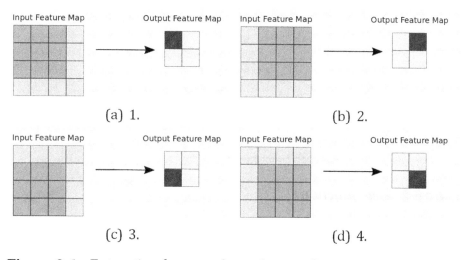

(a) 1.   (b) 2.

(c) 3.   (d) 4.

Figure 6-1. *Extracting features through convolution*

2. A Relu layer is introduced after the convolution layer in order to introduce nonlinearity into the model. The Relu function is defined as $f(x)=max(0,x)$ and is given by Figure 6-2.

3. After application of the Relu activation function, pooling is applied to downsample the convolved feature. Downsampling needs to be done by keeping the most critical feature information while at the same time reducing the dimensions as much as possible. Generally max pooling is used where tiles are extracted from the feature map and the maximum value is kept from each tile while discarding all the other values.

231

At the end of the CNN layer, one or more layers of fully connected layers are needed to perform the classification based on the features extracted by the convolutions.

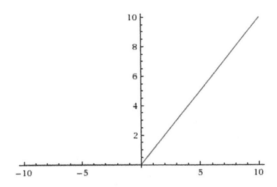

Figure 6-2. *Relu function*

6.1.2 Rust and Torch

As seen in Chapter2, we can use the torch deep learning library,[1] which many developers know as pytorch. Since in the case of computer vision, deep learning has been shown to have the best results, we will be focusing mostly on creating and using deep neural models. Currently torch is the most usable in terms of creating and using deep neural models in Rust; and hence in this chapter, we will mostly be working in the torch paradigm.

6.1.3 Torch Dataset

We will now create a Rust project for image classification. First create a directory chapter 6 and inside it we can create a `pytorch-image-classification` project (Listing 6-1).

---

[1]https://pytorch.org/.

Listing 6-1. Create image classifi cation project

```
$ cd chapter6
$ cargo new pytorch-image-classification --bin
$ cd pytorch-image-classification
$
```

Since we will be working with the torch dependency, let's download the library files as well and create established references to them so that the tch library is able to find the relevant core dependencies (Listing 6-2).

Listing 6-2. Download mkl and libtorch dependencies

```
$ wget https://github.com/intel/mkl-dnn/releases/download/
  v0.19/mklml_mac_2019.0.5.20190502.tgz
$ gunzip -c mklml_mac_2019.0.5.20190502.tgz | tar xvf -
$ wget https://download.pytorch.org/libtorch/cpu/libtorch-
  macos-1.1.0.zip
$ unzip libtorch-macos-1.1.0.zip
$ export LD_LIBRARY_PATH=mklml_mac_2019.0.5.20190502/lib:"$LD_
  LIBRARY_PATH"
$ export LIBTORCH=libtorch
$ export LD_LIBRARY_PATH=${LIBTORCH}/lib:$LD_LIBRARY_PATH
$
```

To build an image classification model, we will be using the Caltech101[2] dataset. To download the files, go to the link that is given in the footnote and download the images (Listing 6-3).

---

[2]http://www.vision.caltech.edu/Image_Datasets/Caltech101/.

Listing 6-3. Download images

```
$ wget http://www.vision.caltech.edu/Image_Datasets\
/Caltech101/101_ObjectCategories.tar.gz
$ gunzip -c 101_ObjectCategories.tar.gz | tar xvf -
$
```

Or you can download using the browser. If you click on the link
`http://www.vision.caltech.edu/Image_Datasets/Caltech101/` in
your browser, you should get a link as shown in Figure 6-3. If downloaded
using the browser, you would still need to uncompress it using the gunzip
command.

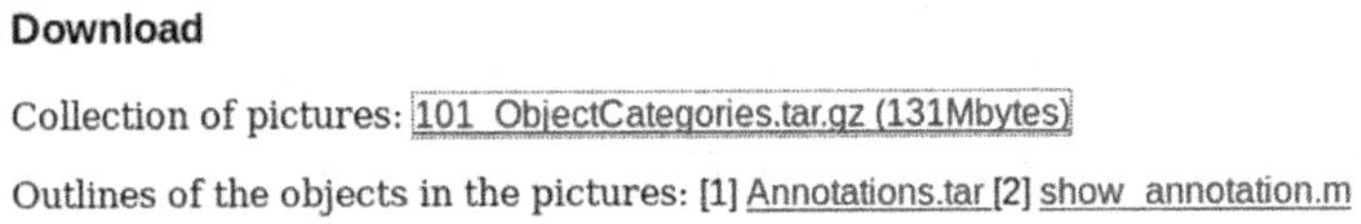

Download

Collection of pictures: 101_ObjectCategories.tar.gz (131Mbytes)

Outlines of the objects in the pictures: [1] Annotations.tar [2] show_annotation.m

Figure 6-3. *Download caltech101*

Similar to `torch-vision`, to work with image datasets, we have the
`vision::Dataset` structure. Let;s take a look at this structure (Listing 6-4).

Listing 6-4. tch-rs/src/vision/dataset.rs

```
#[derive(Debug)]
pub struct Dataset {
  pub train_images: Tensor,
  pub train_labels: Tensor,
  pub test_images: Tensor,
  pub test_labels: Tensor,
  pub labels: i64,
}
```

A handy function in the `tch` crate to load the images and convert
it to the `dataset` mentioned above is through using the function
`tch::vision::imagenet::load_from_dir`. To do that, we will need to

separate the caltech101 files that we have created into training and val folders in the below format. Inside the root folder, we need to have two folders, "train" and "val," and inside those folders we need to have the different images inside the folders with folder names as the labels. Or, in other words, all images with the same label would be in the same folder (Listing 6-5).

Listing 6-5. Dataset directory that we need

```
dataset
 train
    accordion
        image_0001.jpg
        image_0002.jpg
    airplanes
        image_0001.jpg
         image_0060.jpg

         ...
val
    accordion
        image_0036.jpg
    airplanes
        image_0685.jpg

        ...
```

The caltech dataset when uncompressed is in this format (Listing 6-6).

Listing 6-6. Current directory structure

```
101_ObjectCategories
 accordion
     image_0001.jpg
     image_0002.jpg
 airplanes
```

```
image_0001.jpg
image_0002.jpg
```

. . .

```
102 directories, 9145 files
```

Although the directory structure is mostly similar, we still will need to write a small function that will split the dataset into train and validation sets. The function in Listing 6-7 is written for this purpose and is a recursive function where if path is a directory, then we will go deeper or else we will run the testing function on one file and training function on the other files. The variable this_label is there to check if the testing function has already been run on the current label. We will implement the train_fn and test_fn as functions that implement the actual movement. This could have been implemented is the same visit_dir function, but implementing this as separate functions will make this more modular and easily testable.

Listing 6-7. chapter6/pytorch-image-classifi cation/src/main.rs

```rust
use std::io;
use std::fs::{self, DirEntry, copy, create_dir_all};
use std::path::Path;

fn visit_dir(dir: &Path,
             train_fn: &Fn(&DirEntry),
             test_fn: &Fn(&DirEntry)) -> io::Result<()> {
  if dir.is_dir() {
    let mut this_label = String::from("");
    for entry in fs::read_dir(dir)? {
      let entry = entry?;
      let path = entry.path();
      if path.is_dir() {
        visit_dir(&path, train_fn, test_fn)?;
      } else {
```

```
        let full_path: Vec<String> = path.to_str().unwrap()
          .split("/").into_iter()
          .map(|x| x[..].to_string()).collect();
        if this_label == full_path[1] {
          train_fn(&entry); // move the training file
        } else {
          test_fn(&entry); // move the testing file
        }
        // the second entry is the label
        this_label = full_path[1].clone();
      }
    }
  }
  Ok(())
}
```

Now we will need to define the actual file copying function from the source folder to the destination folder. Remember that our destination path should be in the format "root_dataset_foldername > 'train' or 'val' folder based on if the file is a training file or a validation file > label as folder name for the particular image > image file." The function should create all the intermediate directories in case they are not present. This can be implemented using the move_file function in Listing 6-8.

Listing 6-8. chapter6/pytorch-image-classifi cation/src/main.rs

```
use std::fs::{self, DirEntry, copy, create_dir_all};

const DATASET_FOLDER: &str = "dataset";

fn move_file(
        from_path: &DirEntry, to_path: &Path)
        -> io::Result<()> {
  let root_folder = Path::new(DATASET_FOLDER);
```

```
let second_order = root_folder.join(to_path);
let full_path: Vec<String> = from_path.path().to_str().
unwrap()
    .split("/").into_iter().map(|x| x[..].to_string()).
    collect();
let label = full_path[1].clone();
let third_order = second_order.join(label);
create_dir_all(&third_order)?;
let filename = from_path.file_name();
let to_filename = third_order.join(&filename);
copy(from_path.path(), to_filename)?;
Ok(())
}
```

As we can see in Listing 6-8, root folder is the dataset folder, and to_path is essentially if the file needs to go to the train folder or the validation folder. Next we will split the source path to the full_path variable so that we are able to retrieve the label from the full path. Once all the paths are joined, the function create_dir_all from std::fs standard library is used to create all the directories. The respective file name is parsed from the from_path variable, joined to the to_filename variable. Once we have successfully constructed the full path for the destination path, we pass the destination path: in other words, the to_filename variable to the copy function from std::fs standard library to create the folder structure.

We should now be able to run this code and segregate the dataset to train and val images (Listing 6-9).

Listing 6-9. chapter6/pytorch-image-classifi cation/src/main.rs

```
fn main() {
  let obj_categories = Path::new("101_ObjectCategories");
  let move_to_train = |x: &DirEntry| {
    let to_folder = Path::new("train");
    move_file(&x, &to_folder).unwrap();
  }

  let move_to_test =
    |x: &DirEntry| {
      let to_folder = Path::new("val");
      move_file(&x, &to_folder).unwrap();
    };
    visit_dir(
      &obj_categories, &move_to_train,
      &move_to_test).unwrap();
    println!(
      "files kept in the imagenet format in {}", DATASET_
      FOLDER);
  ... remaining code
```

Run the code upto Listing 6-9 here and you should see that the folders are being moved into the respective folders (Listing 6-10).

Listing 6-10. pytorch-image-classifi cation output for now

```
$ cargo run
warning: clang: warning: -Wl,-rpath=chapter6/pytorch-image-
classification/libtorch/lib: 'linker' input unused [-Wunused-
command-line-argument]
    Finished dev [unoptimized + debuginfo] target(s) in 0.05s
```

```
    Running `target/debug/pytorch-image-classification`
files kept in the imagenet format in dataset
$ ls dataset
train val
```

In the code in Listing 6-10, we create two closures, move_to_train and move_to_test that will pass to move_file function the appropriate destination folder. These closures are then passed to the visit_dir function that will run them as we had already seen. Running the code until now should segregate the folders and arrange them in the imagenet format. We should now be able to use the load_from_dir function, which is a handy function in torch::vision. This will load the directory in the correct format so that tch will be able to act on it. See Listing 6-11.

Listing 6-11. chapter6/pytorch-image-classifi cation/src/main.rs

```rust
use tch::vision::imagenet::load_from_dir;

fn main() {
  // previous code ...

  let image_dataset = load_from_dir(DATASET_FOLDER).unwrap();

  // remaining code ...
}
```

Pytorch and hence tch uses the NCHW format for images. If the image shape is [256, 3, 224, 224], it means that the number of samples is 256, there are 3 channels for each image, and 224x224 is the height and width of the dataset.

6.1.4 CNN Model

Now that our data is in the correct format, we will work on defining the model. The advantage in Rust is that the model can be a simple structure (Listing 6-12).

Listing 6-12. chapter6/pytorch-image-classifi cation/src/main.rs

```
use tch::{Device, Tensor, nn};
use tch::nn::{ModuleT, OptimizerConfig, conv2d, linear};

#[derive(Debug)]
struct CnnNet {
  conv1: nn::Conv2D,
  conv2: nn::Conv2D,
  fc1: nn::Linear,
  fc2: nn::Linear,
}
```

A new method for the struct in Listing 6-12 can be implemented, which creates an instance of CnnNet.

Listing 6-13. chapter6/pytorch-image-classifi cation/src/main.rs

```
const LABELS: i64 = 102;
const W: i64 = 224;
const H: i64 = 224;
const C: i64 = 3;

impl CnnNet {
  fn new(vs: &nn::Path) -> CnnNet {
    let conv1 = conv2d(vs, C, 32, 5, Default::default());
    let conv2 = conv2d(vs, 32, 64, 5, Default::default());
    let fc1 = linear(vs, 1024, 1024, Default::default());
    let fc2 = linear(vs, 1024, LABELS, Default::default());
    CnnNet {
      conv1,
      conv2,
```

```
        fc1,
        fc2,
    }
  }
}
```

In Listing 6-13, conv1 is a convolutional layer with input dimension C, which is a constant of 3 and an output dimension of 32. The input dimension needs to be 3, because we will define conv1 to be the first convolutional layer and in our dataset each image is comprised of 3 channels. Further conv1 can have a kernel size of 5x5, which is given by the fourth parameter. The last parameter is the convolutional config, and we can pass default parameters here. The default parameters can be seen in the conv.rs file in the tch-rs library (Listing 6-14).

Listing 6-14. tch-rs/src/nn/conv.rs

```
impl Default for ConvConfig {
  fn default() -> Self {
    ConvConfig {
        stride: 1,
        padding: 0,
        dilation: 1,
        groups: 1,
        bias: true,
        ws_init: super::Init::KaimingUniform,
        bs_init: super::Init::Const(0.),
    }
  }
}
```

Moving on with the description of Listing 6-8, we will have conv2 similar to conv1 with the appropriate dimensions. These dimensions will be explained soon. We will also create two densenets fc1 and fc2, with the last dimension of fc2 the same as the number of labels defined by the LABELS constant.

In pytorch, the python equivalent, the model is created by subclassing from torch.nn.Module. Similarly, when using tch in Rust, we will need to implement nn::ModuleT trait, which will have the relevant functions that can be used for training the model. See Listing 6-15.

Listing 6-15. chapter6/pytorch-image-classifi cation/src/main.rs

```rust
use tch::{Device, Tensor, nn};

const BATCH_SIZE: i64 = 16;

impl nn::ModuleT for CnnNet {
  fn forward_t(&self, xs: &Tensor, train: bool) -> Tensor {
    xs.view(&[-1, C, H, W])     // out dim: [16, 3, 224, 224]
      .apply(&self.conv1)       // [16, 32, 220, 220]
      .max_pool2d_default(2)    // [16, 32, 110, 110]
      .apply(&self.conv2)       // [16, 64, 106, 106]
      .max_pool2d_default(2)    // [16, 64, 53, 53]
      .view(&[BATCH_SIZE, -1]); // [16, 179776]
      .apply(&self.fc1)         // [16, 1024]
      .relu()                   // [16, 1024]
      .dropout_(0.5, train)     // [16, 1024]
      .apply(&self.fc2)         // [16, 102]
  }
}
```

In the code in Listing 6-15, the code, along with the changing dimensions with each tensor transformation, is shown. The dimensions of the input tensor xs changes from [16, 3, 224, 224] to [16,102] dimensions, which is (batch_size, outputs) as the transformations are applied. When you are executing this code, you can verify if the dimensions match in your case as well.

Once the model is created, we should be able to train it on the dataset. The training process for the torch model will generally be similar. We initialize the model instance, we define the loss function, and then we initialize the optimizer. We can now change the return name to be of type Fallible as well, which is a standard in the tch repo. This can be seen in the code given in Listing 6-16.

Listing 6-16. chapter6/pytorch-image-classifi cation/src/main.rs

```
use failure;

fn main() -> failure::Fallible<()> {
  // previous code ...

  let image_dataset = load_from_dir(dataset_folder).unwrap();
  let vs = nn::varstore::new(device::cuda_if_available());
  let opt = nn::adam::default().build(&vs, 1e-4)?;
  let net = cnnnet::new(&vs.root());
  for epoch in 1..100 {
    for (bimages, blabels)
        in image_dataset.train_iter(batch_size)
        .shuffle().to_device(vs.device()) {
      let loss = net
        .forward_t(&bimages, true)
        .cross_entropy_for_logits(&blabels);
      opt.backward_step(&loss);
    }
    let test_accuracy =
      net.batch_accuracy_for_logits(&image_dataset.test_images,
                                    &image_dataset.test_labels,
                                    vs.device(), 1024);
```

```
    println!("epoch: {:4} test acc: {:5.2}%",
            epoch, 100. * test_accuracy,);
}

// remaining code ...

Ok(())
}
```

In Listing 6-16, vs is used to store the common configurations to be used in different stages of the training process. As stated earlier, the model is initialized with the variable net. For 100 epochs, batches are taken from the dataset for training. Since the training variable is an instance of the Dataset struct, the bimages and blabels are neatly retrieved and in this case they are of the shape [32, 3, 244, 244] for the images and [32] for the labels. The forward_t method will pass the tensors through the whole architecture defined previously and opt.backward_step will compute the gradients. After the completion of 100 epochs, we should have our model trained.

You should now be able to run the package using cargo run. Once run, check if the trained model.ot file gets generated.

Note   Since all this earlier code uses a lot of computing resources, your laptop may not be able to run the code. If that happens, you can try reducing the number of labels that you are predicting. In that case, do change the LABELS parameter as well and make it the same as the number of labels that you are predicting. You can also try reducing the BATCH_SIZE as well.

6.1.5 Model Building and Debugging

When building the models, it is very important to keep track of the shapes for the different tensors that are built as the data flows. Otherwise, getting errors such as those shown in Listing 6-17 are quite common.

Listing 6-17. Possible size mismatch error when training

```
thread 'main' panicked at 'called `Result::unwrap()` on an
`Err`
value: TorchError { c_error: "size mismatch,
m1: [2809 x 1024], m2: [179776 x 1024] at
/pytorch/aten/src/TH/generic/THTensorMath.cpp:961" }',
src/libcore/result.rs:997:5
```

Or it could be something like what is shown in Listing 6-18.

Listing 6-18. Possible batch size failure when training

```
thread 'main' panicked at 'called `Result::unwrap()` on an
`Err`
value: TorchError { c_error:
"Assertion `THTensor_sizeLegacyNoScalars(target, 0) == batch_
size\' failed.
at /pytorch/aten/src/THNN/generic/ClassNLLCriterion.c:84" }',
src/libcore/result.rs:997:5
```

In those scenarios, we would need to debug the net. One way of debugging is that we can transfer the app to a python application and run the app through there, printing the shape throughout as we go. See Listing 6-19.

Listing 6-19. Ex code in pytorch

```python
import torch.nn as nn
import torch.nn.functional as F

class Net(nn.Module):
    def __init__(self):
        super(Net, self).__init__()
        self.conv1 = nn.Conv2d(3, 32, 5)
        self.pool1 = nn.MaxPool2d(2, 2)
        self.fc1 = nn.Linear(1600, 1024)
        self.fc2 = nn.Linear(1024, 10)

    def forward(self, x):
        x = self.conv1(x)
        print(x.shape)
        x = self.pool1(x)
        print(x.shape)
        x = F.relu(self.fc1(x))
        print(x.shape)
        x = F.dropout(x, training=True)
        print(x.shape)
        x = self.fc2(x)
        print(x.shape)
        return x

net = Net()
```

The above logic can be utilized in case of Rust as well. We can transform the previous forward function to the above way of writing and plug in println statements as we go (Listing 6-20).

Listing 6-20. Put print statements in the convnet for good debugging

```
impl nn::ModuleT for CnnNet {
    fn forward_t(&self, xs: &Tensor, train: bool) -> Tensor {
        let xs_prime = xs.view(&[-1, C, H, W]);
        println!("{:?}", xs_prime.size());
        let xs_prime = xs_prime.apply(&self.conv1);
        println!("{:?}", xs_prime.size());
        let xs_prime = xs_prime.max_pool2d_default(2);
        println!("{:?}", xs_prime.size());
        let xs_prime = xs_prime.apply(&self.conv2);
        println!("{:?}", xs_prime.size());
        let xs_prime = xs_prime.max_pool2d_default(2);
        println!("{:?}", xs_prime.size());
        let xs_prime = xs_prime.view(&[BATCH_SIZE, -1]);
        println!("{:?}", xs_prime.size());
        let xs_prime = xs_prime.apply(&self.fc1);
        println!("{:?}", xs_prime.size());
        let mut xs_prime = xs_prime.relu();
        println!("{:?}", xs_prime.size());
        let xs_prime = xs_prime.dropout_(0.5, train);
        println!("{:?}", xs_prime.size());
        let xs_prime = xs_prime.apply(&self.fc2);
        println!("{:?}", xs_prime.size());
        xs_prime
    }
}
```

This should be able to print the dimensions for the different layers when running through the network.

6.1.6 Pretrained Models

Generally, once the models are trained, they need to be shipped to a production environment where the model inference happens. In the tch framework, the trained model can be saved as an .ot file and then loaded and served in the inference environment. A variable store can be used to save the model (Listing 6-21).

Listing 6-21. Model saving

```
fn main() {
  // previous code with model training ...

  vs.save("model.ot")?;
}
```

This should save the model in the current directory once we run the training using `cargo run`. Once the model is saved, we can then load the model in the inference app and run predictions on it. Since each model architecture is different, there is no boilerplate functioning code that will work in all scenarios. The developer would need access to each architecture to be able to make predictions.

The `tch/examples/pretrained` models have examples of how to load a pretrained model in case they belong to one of the standard architectures. The standards architectures that are possible to be loaded are alexnet, densenet, imagenet, resnet, squeezenet, and VGG16. Since the code is pretty standard, we will go with the pretrained model example that is shown in the tch repo. Loading the models can be seen using the code shown in Listing 6-22.

Listing 6-22. `https://github.com/LaurentMazare/tch-rs/blob/` `master/examples/pretrained-models/main.rs`

```
use tch::vision::{
        alexnet, densenet, imagenet,
        inception, resnet, squeezenet, vgg};
```

We can load the image and resize it to the imagenet dimension of 224x224, which is a convention that is used when working with this package (Listing 6-23).

Listing 6-23. `https://github.com/LaurentMazare/tch-rs/blob/` `master/examples/pretrained-models/main.rs`

```
extern crate failure;
extern crate tch;
use tch::nn::ModuleT;
use tch::vision::{
    alexnet, densenet, efficientnet, imagenet, inception,
    mobilenet, resnet, squeezenet, vgg,
};

pub fn main() -> failure::Fallible<()> {
  let args: Vec<_> = std::env::args().collect();
    let (weights, image) = match args.as_slice() {
        [_, w, i] => (std::path::Path::new(w), i.to_owned()),
        _ => bail!("usage: main resnet18.ot image.jpg"),
    };

  let image = imagenet::load_image_and_resize224(image)?;

  // remaining code ...
}
```

Now let's try working with the resnet18 model. We would need to have a way to encode the architecture in our deep network, which needs to be an instance of the ModuleT struct. Hence we create a VarStore variable and create a resnet18 net instance (Listing 6-24).

Listing 6-24. `https://github.com/LaurentMazare/tch-rs/blob/` `master/examples/pretrained-models/main.rs`

```
pub fn main() -> failure::Fallible<()> {
  // previous code ...

  let mut vs = tch::nn::VarStore::new(tch::Device::Cpu);
  let net: Box<dyn ModuleT> = Box::new(resnet::resnet18
                                   (&vs.root(),imagenet::
                                   CLASS_COUNT)),

  // remaining code ...
```

Now that the model basic architecture is initialized, we should be able to load the weights and create the trained model (Listing 6-25).

Listing 6-25. `https://github.com/LaurentMazare/tch-rs/blob/` `master/examples/pretrained-models/main.rs`

```
pub fn main() -> failure::Fallible<()> {
  // previous code ...

  vs.load(weights)?;

  // remaining code ...
}
```

Once we are able to create the model, we should now be able to run predictions on our model. See Listing 6-26.

Listing 6-26. https://github.com/LaurentMazare/tch-rs/blob/
master/examples/pretrained-models/main.rs

```
pub fn main() -> failure::Fallible<()> {
  // previous code ...

  // Apply the forward pass of the model to get the logits.
  let output = net
    .forward_t(&image.unsqueeze(0), /*train=*/ false)
    .softmax(-1); // Convert to probability.

  // Print the top 5 categories for this image.
  for (probability, class) in imagenet::top(&output, 5).iter() {
    println!("{:50} {:5.2}%", class, 100.0 * probability)
  }

  // remaining code ...
}
```

Similar to loading a saved model for standard and common architecture, for the models that we have trained and saved, we will need access to the original model architecture struct that was created to create the model. So, we will create a path to the model file and the image file. (Listing 6-27)

Listing 6-27. chapter6/pytorch-image-classifi cation/src/main.rs

```
fn main() -> failure::Fallible<()> {
  // previous model building code ...

  let weights = Path::new("model.ot");
  let image = "image.jpg";

  // remaining code ...
```

After that, the code that we will see is similar to Listing 6-27 except that in this case we will be creating a model from the CnnNet that we had created before. See Listing 6-28.

Listing 6-28. chapter6/pytorch-image-classifi cation/src/main.rs

```
pub fn main() -> failure::Fallible<()> {
  // previous code ...

  // Load the image file and resize it to the usual imagenet
  dimension of 224x224.
  let image = load_image_and_resize224(image)?;

  // Create the model and load the weights from the file.
  let mut vs = tch::nn::VarStore::new(tch::Device::Cpu);
  let net: Box<dyn ModuleT> = Box::new(CnnNet::new(&vs.
  root()));
  vs.load(weights)?;

  // Apply the forward pass of the model to get the logits.
  let output = net
    .forward_t(&image.unsqueeze(0), /*train=*/ false)
    .softmax(-1); // Convert to probability.

  // Print the top 5 categories for this image.
  for (probability, class) in top(&output, 5).iter() {
    println!("{:50} {:5.2}%", class, 100.0 * probability)
  }

  Ok(())
}
```

6.2 Transfer Learning

Training a sufficiently deep neural network for high accuracy is generally not feasible as neural nets require a huge amount of data, and such a large amount of data may not be present, especially in the early stages of the product. Instead a pretrained conv net that is trained on a large dataset is either used as an initializer or for gathering the set of features from the image. In this case, we will freeze the weights for all the network except that of the final layer. This last fully connected layer is replaced with a new one with random weights, and only this layer is trained. A possible transfer learning architecture is shown in Figure 6-4.

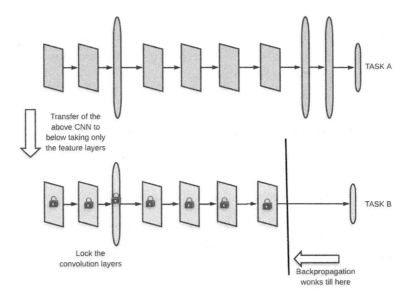

Figure 6-4. *Transfer Learning*

To proceed with this example, we will download a pretrained model from this url: `https://github.com/LaurentMazare/ocaml-torch/ releases/download/v0.1-unstable/resnet18.ot`. See Listing 6-29.

Listing 6-29. Download resnet pretrained model

```
wget https://github.com/LaurentMazare/ocaml-torch/releases/
download/v0.1-unstable/resnet18.ot
```

As the code for this section is also standard on how to load and make predictions, we will follow the example in main repo.

To be able to do this, we will need to load the dataset (Listing 6-30).

Listing 6-30. chapter6/pytorch-image-classifi cation/src/main.rs

```
#[macro_use]
extern crate failure;
extern crate tch;
use tch::nn::{self, OptimizerConfig};
use tch::vision::{imagenet, resnet};

pub fn main() -> failure::Fallible<()> {
    let args: Vec<_> = std::env::args().collect();
    let (weights, dataset_dir) = match args.as_slice() {
        [_, w, d] => (std::path::Path::new(w), d.to_owned()),
        _ => bail!("usage: main resnet18.ot dataset-path"),
    };
    let dataset = imagenet::load_from_dir(dataset_dir)?;

    // remaining code ...
```

Similar to the pretrained network, we will need to create the model and load the weights to the model. As seen before, we can keep the weights in a variable store (Listing 6-31).

Listing 6-31. `https://github.com/LaurentMazare/tch-rs/blob/`
`master/examples/transfer-learning/main.rs`

```
let mut vs = tch::nn::VarStore::new(tch::Device::Cpu);
let net = resnet::resnet18_no_final_layer(&vs.root());
vs.load(weights)?;
```

We will need to store the output of the model in such a way that we are
able to create the resulting vectors and tell pytorch that we don't need to store
the graph as we are not going to compute the gradients. See Listing 6-32.

Listing 6-32. `https://github.com/LaurentMazare/tch-rs/blob/`
`master/examples/transfer-learning/main.rs`

```
let train_images = tch::no_grad(
  || dataset.train_images.apply_t(&net, false));
let test_images = tch::no_grad(
  || dataset.test_images.apply_t(&net, false));
```

6.2.1 Training

We can now append a small linear layer on top of the extracted layers and
train only the linear part of the layer. This way, less of our data will be used
to effectively train only a small part of the net instead of trying to train on
the whole model. We create the model and a variable store is used to store
the trainable variables (Listing 6-33).

Listing 6-33. `https://github.com/LaurentMazare/tch-rs/blob/`
`master/examples/transfer-learning/main.rs`

```
let vs = tch::nn::VarStore::new(tch::Device::Cpu);
let linear = nn::linear(vs.root(), 512, dataset.labels,
Default::default());
```

Similar to the previous code, we can use adam to minimize the cross entropy loss in the classification task. Hence we will need to create an optimizer and iterate on the training dataset. After each epoch, the accuracy is computed on the training set and printed (Listing 6-34).

Listing 6-34. `https://github.com/LaurentMazare/tch-rs/blob/master/examples/transfer-learning/main.rs`

```
let optimizer = nn::Adam::default().build(&vs, 1e-4)?;
for epoch_idx in 1..1001 {
  let predicted = train_images.apply(&linear);
  let loss = predicted.cross_entropy_for_logits(&dataset.train_
  labels);
  optimizer.backward_step(&loss);

  let test_accuracy = test_images
    .apply(&linear)
    .accuracy_for_logits(&dataset.test_labels);
  println!("{} {:.2}%", epoch_idx, 100. * f64::from(test_
  accuracy));
}
```

Once the training is done, we can save the file using `vs.save("model.ot")?;` which can be again reused in a later training cycle when the next batch of images is ready.

6.2.2 Neural Style Transfer

Neural style transfer is a technique used to generate images in the style of another image. The neural style algorithm takes a content-image and a style image as input and returns the content image transformed in the style of the style image. This is done by creating two distances: one for the content and another for the image. A new image is created such that the content distance from the content image and the style distance from the style image is minimized.

Content Loss   To create such a transformed image, we need a way to control the amount of content image that ends up in the optimized output image. For this, a way to determine the content loss needs to be determined. The main challenge in the content loss is figuring out a way to extract only the content features of an image, and not the style of the image. The solution to this as per the Artistic Style paper [2] is that using the feature maps of the different convolutional layers should solve this problem. Trained convnets learn to represent different parts of the image; the initial layers capture the rough, underlying pattern; and the final layers capture the distinct features. It's the intermediate layers that capture the spatial characteristics of the image, which is what is needed here.

Thus, in this case the content loss is calculated by just computing the mean-square error on some of the upper layers. This error helps in ensuring that the extracted images happen at the same on-screen places between the content and current images (Listing 6-35).

Listing 6-35. `https://github.com/LaurentMazare/tch-rs/blob/ master/examples/neural-style-transfer/main.rs`

```
let content_loss: Tensor = CONTENT_INDEXES
  .iter()
  .map(
    |&i| input_layers[i].mse_loss(&content_layers[i], 1))
  .sum();
```

Style loss   The style loss is responsible for incorporating the style of image in the final image. According to the paper, the style of an image can be encoded using a gram matrix of the image. The gram matrix is made by computing the dot product of the flattened style features and themselves (Listing 6-36).

Listing 6-36. `https://github.com/LaurentMazare/tch-rs/blob/`
`master/examples/neural-style-transfer/main.rs`

```
fn gram_matrix(m: &Tensor) -> Tensor {
  let (a, b, c, d) = m.size4().unwrap();
  let m = m.view(&[a * b, c * d]);
  let g = m.matmul(&m.tr());
  g / (a * b * c * d)
}

fn style_loss(m1: &Tensor, m2: &Tensor) -> Tensor {
  gram_matrix(m1).mse_loss(&gram_matrix(m2), 1)
}

pub fn main() -> failure::Fallible<()> {
  // previous code ...

  let style_loss: Tensor = STYLE_INDEXES
    .iter()
    .map(
      |&i| style_loss(&input_layers[i], &style_layers[i]))
    .sum();

  // remaining code ...

}
```

Observe that the style of an image is not dependent on the pixel values but the relationship between the pixel values. The gram matrix delocalizes all the information of the style image, such as texture, shape, and weights, and then the dot product is taken. This results in features that co-occur more get greater weightage, and features that do not have similarity across the style image do not get weightage.

Now that we have the basic idea on how style transfer works, let's look at the full implementation. To be able to do neural style transfer, we will need to load the content image and the style image. These will need to be torch tensors and hence we can use the imagenet module from `tch::vision` to load the images (Listing 6-37).

Listing 6-37. `https://github.com/LaurentMazare/tch-rs/blob/master/examples/neural-style-transfer/main.rs`

```
pub fn main() -> failure::Fallible<()> {
  // previous code ...

  let style_img = imagenet::load_image(style_img)?
    .unsqueeze(0).to_device(device);
  let content_img = imagenet::load_image(content_img)?
    .unsqueeze(0).to_device(device);

  // remaining code ...

}
```

We will also need to load a pretrained model. We can use a 19-layer VGG network like the one used in the paper. This is another hyperparameter that you can experiment with and see if there can be another model that would provide a better feature for the content image. Download the model by going to this link or using your favorite download command (Listing 6-38).

Listing 6-38. Download vgg pretrained model

```
$ wget https://github.com/LaurentMazare/ocaml-torch/releases/
  download/v0.1-unstable/vgg16.ot
$
```

This VGG model consists of two Sequential child modules: features (containing the convolution and pooling layers), and the classifier (containing the fully connected layers). To be able to capture the content from the image, we will need to use the features layer. We will load the weights from the model and freeze the model (Listing 6-39).

Listing 6-39. `https://github.com/LaurentMazare/tch-rs/blob/master/examples/neural-style-transfer/main.rs`

```rust
pub fn main() -> failure::Fallible<()> {
  // previous code ...

  let mut net_vs = tch::nn::VarStore::new(device);
  let net = vgg::vgg16(&net_vs.root(),
                       imagenet::CLASS_COUNT);
  net_vs.load(weights)?;
  net_vs.freeze();

  // remaining code ...
}
```

We now run the model on the style and content images. These calls return a vector of the extracted features for each of the model layers (Listing 6-40).

Listing 6-40. `https://github.com/LaurentMazare/tch-rs/blob/master/examples/neural-style-transfer/main.rs`

```rust
pub fn main() -> failure::Fallible<()> {
  // previous code ...

  let style_layers = net.forward_all_t(
    &style_img, false, Some(max_layer));
  let content_layers = net.forward_all_t(
    &content_img, false, Some(max_layer));

  // remaining code ...
}
```

We will use a gradient optimization method to optimize an image. For that we create a second variable store to hold this image. The initial values for this image will be by copying the content image. This image will be transformed so that the style is similar to the style image. See Listing 6-41.

Listing 6-41. `https://github.com/LaurentMazare/tch-rs/blob/master/examples/neural-style-transfer/main.rs`

```
pub fn main() -> failure::Fallible<()> {
  // previous code ...

  let vs = nn::VarStore::new(device);
  let input_var = vs.root().var_copy("img", &content_img);

  // remaining code ...
}
```

We will now create the optimizer, which will be an instance of the use Adam optimization algorithm (Listing 6-42).

Listing 6-42. `https://github.com/LaurentMazare/tch-rs/blob/master/examples/neural-style-transfer/main.rs`

```
pub fn main() -> failure::Fallible<()> {
  // previous code ...

  let opt = nn::Adam::default().build(&vs, LEARNING_RATE)?;

// remaining code ..}
```

Now in the gradient descent loop cycle, the content and style losses will need to be computed. They will then be summed together, and the aggregate loss will be optimized using the optimizer created in Listing 6-42. For every 1000 epochs, we can compute the current loss and write the current image to file (Listing 6-43).

Listing 6-43. https://github.com/LaurentMazare/tch-rs/blob/
master/examples/neural-style-transfer/main.rs

```rust
pub fn main() -> failure::Fallible<()> {
  // previous code ...

  for step_idx in 1..(1 + TOTAL_STEPS) {
    let input_layers = net.forward_all_t(
      &input_var, false, Some(max_layer));
    let style_loss: Tensor = STYLE_INDEXES
      .iter()
      .map(
        |&i| style_loss(&input_layers[i], &style_layers[i]))
      .sum();
    let content_loss: Tensor = CONTENT_INDEXES
      .iter()
      .map(
        |&i| input_layers[i].mse_loss(&content_layers[i], 1))
      .sum();
    let loss = style_loss * STYLE_WEIGHT + content_loss;
    opt.backward_step(&loss);
    if step_idx % 1000 == 0 {
      println!("{} {}", step_idx, f64::from(loss));
      imagenet::save_image(&input_var, &format!("out{}.jpg",
      step_idx))?;
    }
  }

  // remaining code ...
}
```

We should be able to run the code now. Keep the TOTAL_STEPS 5000 or higher so that we have a high number of images and can choose the image that seems the most coherent yet similar to the target style.

6.3 Tensorflow and Face Detection

In the previous sections, we have mostly worked on torch and the corresponding Rust framework tch. In this section we will be using tensorflow and look at an application for face detection. The challenge is to take an image, identify the faces, and create a new image with boxes drawn around the faces. As seen in Chapter 2, just running the inference in tensorflow is easier than training a model, and hence we will mostly be working toward model inference in this case.

Create the face-detection-tf package for detecting faces and boxing the areas around them. See Listing 6-44.

Listing 6-44. Create face-detection-tf Rust package

```
$ cd chapter6
$ cargo new face-detection-tf --bin
$ cd face-detection-tf
$
```

Download the following trained model. This model will be used to predict an image. Figure 6-5 is the predicted image with the faces boxed.[3] See Listing 6-45.

Listing 6-45. Download mtcnn pretrained model

```
$ wget https://github.com/blaueck/tf-mtcnn/raw/master/mtcnn.pb
$
```

---

[3]Original Image Source http://sannadullaway.com/7xdocsffg0d917t2gqkbqpqy z36qgx.

Or you can download from this url: `https://github.com/blaueck/tf-mtcnn/blob/master/mtcnn.pb`.

Now let's talk about the dependencies. We will need the tensorflow crate to read the model and make the inferences on an image using the model. Structopt crate will be used to parse the command line and make the input output parsing better. Apart from these we will need the image crate to read, write, and make changes in the image. The crate imageproc help in the related image processing operations. Hence we should have a `Cargo.toml` similar to the one shown in Listing 6-46.

Listing 6-46. chapter6/face-detection-tf/src/main.rs

```
[package]
name = "face-detection-tf"
version = "0.1.0"
edition = "2018"

[dependencies]
tensorflow = "0.13.0"
structopt = "0.2.15"
image = "0.21.1"
imageproc = "0.18.0"
```

We can now start with defining the command line. Our app can have the image file name as one of the inputs and an output file name as the file that would need to be created with the faces identified. See Listing 6-47.

Listing 6-47. chapter6/face-detection-tf/src/main.rs

```
use std::path::PathBuf;
use structopt::StructOpt;

#[derive(Debug, StructOpt)]
#[structopt(name = "face-detection-tf", about = "Face
Identification")]
```

```
struct Opt {
    #[structopt(short = "i", long = "input", parse(from_os_
    str))]
    input: PathBuf,

    #[structopt(short = "o", long = "output", parse(from_os_
    str))]
    output: PathBuf
}
```

Creating the above StructOpt struct means that when we plug in let opt = Opt::from_args(); in the main function, we will be able to pass command-line arguments as in ./face-detection-tf -i inputfilename.jpg -o outputfilename.jpg or pass --input and --output as the command-line arguments. To run this check if the command-line arguments run now, we can put the following lines in the main function of the code and check if the command lines are being parsed correctly. See Listing 6-48.

Listing 6-48. chapter6/face-detection-tf/src/main.rs

```
use std::error::Error;
fn main() -> Result<(), Box<dyn Error>> {
    let opt = Opt::from_args();
    // since this is plumbing we will take
    // care that we dont move over ownership of the
    // opt objects using to_owned. In the final code
    // you probably will not need this code.
    println!("{:?}",
      (opt.input.to_owned(), opt.output.to_owned()));
    Ok(())
}
```

Now running cargo run should give us the correct outputs (Listing 6-49).

Listing 6-49. chapter6/face-detection-tf/src/main.rs

```
$ cargo run -- -i input_image.jpg -o out.jpg
   Compiling face-detection-tf v0.1.0 (path to face-detection-tf)

    Finished dev [unoptimized + debuginfo] target(s) in 28.24s
     Running `target/debug/face-detection-tf -i input_image.jpg
     -o out.jpg`
("input_image.jpg", "out.jpg")
$
```

We can now see if we are able to load the model. We will first load the model as a byte array, and then we will try to infer the graph from the byte array. If the model file is corrupted, this should give an error in this case (Listing 6-50).

Listing 6-50. chapter6/face-detection-tf/src/main.rs

```
use tensorflow::Graph;
use tensorflow::ImportGraphDefOptions;
use tensorflow::{
  Session, SessionOptions, SessionRunArgs, Tensor};

fn main() -> Result<(), Box<dyn Error>> {
  ... prev code ...
  let model = include_bytes!("../mtcnn.pb");
  let mut graph = Graph::new();
  graph.import_graph_def(
    &*model, &ImportGraphDefOptions::new())?;
  let session = Session::new(
    &SessionOptions::new(), &graph)?;

  .. rem code ...
```

If the graph is loaded successfully, we should be able to create a session with this graph.

Now if we check the model from which the graph is created, we can see that there are some variables that need to be set before running the model as seen in this line for the model https://github.com/blaueck/tf-mtcnn/blob/master/mtcnn.py#L9. These variables are min_size, factor, and thresholds. We will need to assign those variables in our session before running the session. The values that are assigned in Listing 6-51 are the same as the ones that had been used for the model, but you play around with these values and see if there are better alternatives.

Listing 6-51. chapter6/face-detection-tf/src/main.rs

```
fn main() -> Result<(), Box<dyn Error>> {
  ... previous code ...

  let min_size = Tensor::new(
    &[]).with_values(&[40f32])?;
  let thresholds = Tensor::new(&[3]).with_values(
    &[0.6f32, 0.7f32, 0.7f32])?;
  let factor = Tensor::new(&[]).with_values(&[0.709f32])?;

  let mut args = SessionRunArgs::new();

  //Load our parameters for the model
  args.add_feed(
    &graph.operation_by_name_required("min_size")?,
    0, &min_size);
  args.add_feed(
    &graph.operation_by_name_required("thresholds")?,
    0, &thresholds);
  args.add_feed(
    &graph.operation_by_name_required("factor")?,
    0, &factor);
```

```
... remaining code ...
```

One other variable that needs to be passed, and the most important one that will enable the graph to recognise the faces, is the input image tensor. So we will open the input image using the image crate. Once the image is opened, we will read the values and store in a tensor in BGR format.[4] See Listing 6-52.

Listing 6-52. chapter6/face-detection-tf/src/main.rs

```rust
use image::GenericImageView;

fn get_input_image_tensor(
        opt: &Opt)
        -> Result<Tensor<f32>, Box<dyn Error>> {
  let input_image = image::open(&opt.input)?;

  let mut flattened: Vec<f32> = Vec::new();
  for (_x, _y, rgb) in input_image.pixels() {
    flattened.push(rgb[2] as f32);
    flattened.push(rgb[1] as f32);
    flattened.push(rgb[0] as f32);
  }
  let input = Tensor::new(
    &[input_image.height() as u64, input_image.width()
    as u64, 3])
    .with_values(&flattened)?;

  Ok(input)
}
```

---

[4]We will need to store in BGR instead of RGB because the model was created to support opencv. Go to this link for further information: https://stackoverflow.com/a/33787594/5417164.

So now as seen in Listing 6-53, we should be able to load the image to the graph as the input.

Listing 6-53. chapter6/face-detection-tf/src/main.rs

```
fn main() -> Result<(), Box<dyn Error>> {
  ... previous code ...

  let input = get_input_image_tensor(&opt)?;
  args.add_feed(
    &graph.operation_by_name_required("input")?, 0, &input);

  ... remaining code ...
}
```

Now that the inputs are taken care of, we need to create endpoints for the outputs. This will enable us to fetch the results when the session is run (Listing 6-54).

Listing 6-54. chapter6/face-detection-tf/src/main.rs

```
fn main() -> Result<(), Box<dyn Error>> {
  ... previous code ...

  let bbox = args.request_fetch(
    &graph.operation_by_name_required("box")?, 0);
  let prob = args.request_fetch(
    &graph.operation_by_name_required("prob")?, 0);

  session.run(&mut args)?;

  let bbox_res: Tensor<f32> = args.fetch(bbox)?;
  let prob_res: Tensor<f32> = args.fetch(prob)?;
```

```
println!("{:?}", bbox_res.dims());
println!("{:?}", prob_res.dims());

... remaining code ...
}
```

Executing the code in Listing 6-54 until now would print the dimensions of bbox_res and prob_res (Listing 6-55).

Listing 6-55. face-detection-tf output

```
$ cargo run -- --input Solvay_conference_1927.jpg --output
output.jpg
    Finished dev [unoptimized + debuginfo] target(s) in 0.06s
     Running `target/debug/face-detection-tf --input Solvay_
    conference_1927.jpg --output output.jpg`
("Solvay_conference_1927.jpg", "output.jpg")
2019-08-25 16:30:59.485045: I tensorflow/core/platform/cpu_
feature_guard.cc:141] Your CPU supports instructions that this
TensorFlow binary was not compiled to use: SSE4.2 AVX AVX2 FMA
[30, 4]
[30]
$
```

As we can see, the output dimensions for bbox_res is [30, 4] and prob_res is [30].

Now we will need to parse the outputs and store in a struct so that we are able to draw the bounding boxes. For this reason, we will create a BBox struct so that we are able to store the outputs. See Listing 6-56.

Listing 6-56. chapter6/face-detection-tf/src/main.rs

```
#[derive(Copy, Clone, Debug)]
pub struct BBox {
  pub x1: f32,
  pub y1: f32,
  pub x2: f32,
  pub y2: f32,
  pub prob: f32,
}
```

We should now be able to store the results as a Vec<BBox> (Listing 6-57).

Listing 6-57. chapter6/face-detection-tf/src/main.rs

```
fn main() -> Result<(), Box<dyn Error>> {
  ... previous code ...

  let bboxes: Vec<_> = bbox_res
    .chunks_exact(4)
    .zip(prob_res.iter()) // combine with the probability
    outputs
    .map(|(bbox, &prob)| BBox {
      y1: bbox[0], x1: bbox[1],
      y2: bbox[2], x2: bbox[3],
      prob,
    }).collect();

  ... remaining code ...
}
```

We should now be able to draw the rectangles around the faces for the input figure. For that, we will create a new image, called output_image and which of course should be mutable, with the input image and for all the bboxes we will draw rectangles at x1 and y1 with sizes (bbox.x2 - bbox.x1) and (bbox.y2 - bbox.y1). Once the images have been updated with rectangles, we can save the changed image (Listing 6-58).

Listing 6-58. chapter6/face-detection-tf/src/main.rs

```
use image;
use imageproc;
use imageproc::rect::Rect;
use imageproc::drawing::draw_hollow_rect_mut;

const LINE_COLOUR: Rgba<u8> = Rgba {
    data: [0, 255, 0, 0],
};

fn main() -> Result<(), Box<dyn Error>> {
  ... previous code ...

  let mut output_image = image::open(&opt.input)?;

  for bbox in bboxes {
    let rect = Rect::at(bbox.x1 as i32, bbox.y1 as i32)
      .of_size((bbox.x2 - bbox.x1) as u32, (bbox.y2 - bbox.y1)
      as u32);
    draw_hollow_rect_mut(&mut output_image, rect, LINE_COLOUR);
  }
  output_image.save(&opt.output)?;

  Ok(())
}
```

Figure 6-5. *Famous Scientists*

Download the Solvay_conference_1927.jpg file from this url:
http://sannadullaway.com/7xdocsffg0d917t2gqkbqpqyz36qgx if not
already done, and keep in the face-detection-tf root directory. We should
then be able to pass this file to the code and build the resulting image,
which we can call output.jpg (Listing 6-59).

Listing 6-59. face-detection-tf output

```
cargo run -- --input Solvay_conference_1927.jpg --output output.jpg
    Finished dev [unoptimized + debuginfo] target(s) in 0.07s
    Running `target/debug/face-detection-tf --input Solvay_
    conference_1927.jpg --output output.jpg`
("Solvay_conference_1927.jpg", "output.jpg")
2019-08-25 14:31:02.256970: I tensorflow/core/platform/cpu_
feature_guard.cc:141] Your CPU supports instructions that this
TensorFlow binary was not compiled to use: SSE4.2 AVX AVX2 FMA
[30, 4]
[30]
```

```
BBox Length: 30, Bboxes:[
    BBox {
        x1: 1031.1285,
        y1: 333.9382,
        x2: 1082.9902,
        y2: 397.1919,
        prob: 1.0,
    },
    ... printing all the boxes ...
    BBox {
        x1: 207.15288,
        y1: 357.41956,
        x2: 241.75725,
        y2: 403.062,
        prob: 0.89407635,
    },
]
```

Check the folder and we should see the output.jpg file being formed. Opening the file should give us the image that is shown in the start of this section.

6.4 Conclusion

This chapter introduced us to different interesting applications in Computer Vision. The chapter started with image classification and how to build an image classifier using tch library, which allows easy usage of the pytorch deep learning library. Then the chapter went on with using pretrained networks for image classification when the number of samples in the dataset are small. The next application for discussion was Neural Style transfer, and an application was created that took in a content image and a style image and transformed the content image in the style of the

275

style image. The next application in the chapter was how Generative Neural Networks work and creating a fake facial creator app using the tch library. After mostly working in the tch library, the chapter also included an application in identifying faces in an image using the tensorflow library.

In the next chapter, you will learn about other different needs in machine learning such as statistical analysis in Rust and recommender systems.

6.5 Bibliography

[1]   cetra3, "Face Detection with Tensorflow Rust", [Online; last checked 18 nov 2019]. 2019. URL: https://cetra3.github.io/blog/face-detection-with-tensorflow-rust/

[2]   Leon A. Gatys, Alexander S. Ecker, and Matthias Bethge. "A Neural Algorithm of Artistic Style." In: *CoRR* abs/1508.06576 (2015). arXiv: 1508.06576. URL: http://arxiv.org/abs/1508.06576.

[3]   cetra3. *Face Detection with Tensorflow Rust.*

[4]   Greg Surma. *Style Transfer - Styling Images with Convolutional NeuralNetworks.* Ed. by Towards Data Science. [Online; last checked 19 May 2019]. 2014. URL: https://towardsdatascience.com/style-transferstyling-images-with-convolutional-neural-networks-7d215b58f461.

[5]   Tanish Baranwal. *Neural Style Transfer Using Tensorflow 2.0.* Ed. by Towards Data Science. [Online; last checked 19 May 2019]. 2019. URL: https://towardsdatascience.com/neural-style-transfer-23a3fb4c6a9e.

[6]   Andrej Karpathy. *Convolutional Neural Networks (CNNs / ConvNets).*

## CHAPTER 7

Machine Learning Domains

In the previous two chapters, we took a look at how machine learning is approached in the two major domains of NLP and computer vision. These domains cover the major breakthroughs of machine learning, and the state of the art is continually being pushed forward in these domains. But a lot of machine learning and high performance computing come outside these domains as well. In this chapter, we will take a look at some of these domains and how Rust can help in creating applications in these domains. We shall start with Statistical Analysis where we will compute Z-scores for different diseases in a genetic dataset. Then we will move on to understanding how high performance code can be written using SIMD and BLAS libraries in Rust. Finally, we will build a good books recommender in Rust.

By the end of this chapter, you should have a fair understanding of how machine learning applications can be built in different domains.

7.1 Statistical Analysis

As data scientists and machine learning engineers, we will need to perform a lot of statistical analysis on different types of data. The underlying assumption of this section is that if we are able to parse the data and store different measurements of a variable in an ndarray matrix, we should be able to perform statistical analysis on them. Many programming languages

© Joydeep Bhattacharjee 2020
J. Bhattacharjee, *Practical Machine Learning with Rust*,
https://doi.org/10.1007/978-1-4842-5121-8_7

that have machine learning libraries implement the ndarray matrix type. An ndarray is a multidimensional container of items of the same type and fixed size. The number of dimensions and items in an array is defined by its shape. In Rust the ndarray type is implemented in the ndarray crate. The ndarray crate has been discussed in detail in Chapter 4.

To explore how we can use Rust to compute different statistical parameters, we create a Rust project, named statistics, where we will be writing the code. Change directory to chapter7 in a terminal and create the Rust binary package statistics (Listing 7-1).

Listing 7-1. Rust binary package statistics creation

```
$ cd chapter7
$ cargo new statistics --bin && cd statistics
```

We should now have the main.rs file and the Cargo.toml file. Add the dependencies to the toml file. Since in this case we are trying to compute different statistics for a dataset, we will be needing the ndarray crate in the dependencies (Listing 7-2).

Listing 7-2. chapter7/statistics/Cargo.toml

```
[package]
name = "statistics"
version = "0.1.0"
edition = "2018"

[dependencies]
ndarray = "0.12.1"
```

Now let's move on to the dataset. As an example of how Rust enables us to perform statistical analysis on raw data, we can perform a simple differential expression analysis on a gene expression dataset. We will use false discovery rates to provide interpretable results when conducting an analysis that involves large-scale multiple hypothesis testing. Note the format of the dataset and how we will be reading the dataset to create the vectors of our choice.

We can download the raw data from this url" {`https://www.ncbi.nlm.nih.gov/sites/GDSbrowser?acc=GDS1615`}. See Listing 7-3.

Listing 7-3. Download browser data

```
$ wget ftp://ftp.ncbi.nlm.nih.gov/geo/datasets\
/GDS1nnn/GDS1615/soft/GDS1615_full.soft.gz
$ gunzip GDS1615_full.soft.gz
```

Once we have downloaded the data, we will see that the file is in this format (Listing 7-4).

Listing 7-4. Peep into the file

```
DATABASE = Geo
### DATASET HEADERS Go HERE ...
!dataset_table_begin
## DATASET HEADERS HERE ...
1007_s_at        MIR4640 80.7287 ...
// THE INDIVIDUAL RECORDS ...
AFFX-TrpnX-M_at --Control    1.62238 ...
!dataset_table_end
```

The first 193 lines are the headers of the whole request and other metadata of the data records in the soft file. The records are kept between the lines `!dataset_table_begin` ... `!dataset_table_end`. And the first record after `dataset_table_begin` contains the following data structures:

- GID : A list of gene identifiers of length d

- SID : A list of sample identifiers of length n

- STP : A list of sample descriptions of length d

- X : A dxn array of gene expression values

This dataset contains analysis of peripheral blood mononuclear cells (PBMCs) from 59 Crohn's disease patients and 26 ulcerative colitis (UC) patients. There are 22,283 samples. In this section, we will consider the mean expression levels in the two groups.

Moving on to the code part, we will first need to parse through the whole file. So we create a process_file function and pass the path to the file. Inside the function we can implement the loop to read the lines (Listing 7-5).

Listing 7-5. chapter7/statistics/src/main.rs

```rust
use std::path::Path;
use std::fs::File;
use std::io::{BufRead, BufReader};

fn process_file(filename: &Path) {
  let file = File::open(filename).unwrap();
  for line in BufReader::new(file).lines() {
    let thisline = line?;
    // business logic on the lines implemented.
  }
}

fn main() {
  let filename = Path::new("GDS1615_full.soft");
  process_file(&filename).unwrap();
}
```

Now before moving ahead, let us talk about the dependencies that we would need. Since we will be converting our vectors to ndarray matrices, we will need ndarray. As an option, if you have higher statistics to be measured, we can also bring in the ndarray-stats crate, which gives us a couple of more options when dealing with ndarray vectors.

Now we should be able to read in the headers. The subset_
description comes first, and then in the next line, subset_sample_id are
given as a list. An example of this is shown below in Listing 7-6.

Listing 7-6. Sample subset description and sample id

```
// prev data ..
!subset_description = normal
!subset_sample_id = GSM76115,...
!subset_type = disease state
// subsequent data ...
```

Keeping that in mind, we can write the below code in Listing 7-7.

Listing 7-7. chapter7/statistics/src/main.rs

```
fn process_file(filename: &Path) {
  let mut SIF = HashMap::new();
  let mut subset_description = String::new();
  let mut within_headers = true;
  for line in BufReader::new(file).lines() {
    let thisline = line?;
    let line_split: Vec<String> = thisline
      .split("=")
      .map(|s| s.to_owned()).collect();
    if within_headers {
      if thisline.starts_with("!subset_description") {
        subset_description = line_split[1].trim().to_owned();
      };
      let subset_ids = if thisline.starts_with("!subset_
      sample_id") {
        let subset_ids = line_split[1].split(",");
        let subset_ids = subset_ids.map(|s| s.trim().to_owned());
        subset_ids.collect()
```

```
    } else {
      Vec::new()
    };
    for k in subset_ids {
      SIF.insert(k, subset_description.to_owned());
    }
  }
 }
 // rest of the code ...
}
```

In the code in Listing 7-7, SIF is a mapping from the different subset_ids in the line with subset_sample_id to the subset_description mentioned above them. This needs to go on until dataset_table_begin is reached. See Listing 7-8.

Listing 7-8. chapter7/statistics/src/main.rs

```
'linereading: for line in BufReader::new(file).lines() {
  let thisline = line?;
  let line_split: Vec<String> = thisline.split("=").
  map(|s| s.to_owned()).collect();
  if thisline.starts_with("!dataset_table_begin") {
     within_dataset_table = true;
     within_headers = false;
     continue 'linereading;
  }
// rest of the code ...
}
```

Once the dataset_table_begin is reached, we turn the flag within_headers to false and then jump to the next line. This happens because the for loop is designed that way. That way in the remaining line reads the if within_headers block is not executed anymore.

Once the mapping between subset sample id to subset description (SIF) is done, we should be able to parse through the header, which is the first line after dataset_table_begin. Notice that in the thisline.starts_ with("!dataset_table_begin") block, the within_dataset_table is assigned as true suggesting that we are within the table. We will also have other variable gene_expression_headers to suggest that the header line as passed. We will create a function of the header line as passed. And we will also create a function process_gene_expresssion_data_headers to process the headers for the header line. This function will take the SIF mapping created earlier and will create the column indices for the columns to which the sample descriptions are mapped. See Listing 7-9.

Listing 7-9. chapter7/statistics/src/main.rs

```
fn process_gene_expresssion_data_headers(
    thisline: &String,
    SIF: &HashMap<String, String>)
    -> (Vec<String>, Vec<String>, Vec<usize>) {
  let SID: Vec<String> = thisline.split("\t").map(|s| s.to_
  owned()).collect();
  let indices: Vec<usize> = SID.iter()
              .enumerate()
              .filter(|&(_, x)| x.starts_with("GSM") )
              .map(|(i, _)| i).collect();
  let SID: Vec<String> = indices.iter().map(|&i| SID[i].
  clone()).collect();
  let STP: Vec<String> = SID.iter().map(
      |k| SIF.get(&k.to_string()).unwrap())
      .cloned().collect();
  (SID, STP, indices)
}
```

```
fn process_file(filename: &Path) {
  //prev code ..
  let mut gene_expression_headers = true;
  let mut indices: Vec<usize> = Vec::new();
  let mut gene_expression_measures_vec = Vec::new();
  let mut gene_identifiers_vec = Vec::new();
  let mut SID = Vec::new();
  let mut STP = Vec::new();
  'linereading: for line in BufReader::new(file).lines() {
    let thisline = line?;
    let line_split: Vec<String> = thisline.split("=").map(|s|
    s.to_owned()).collect();
    if within_dataset_table && gene_expression_headers {
      let sid_stp_indices = process_gene_expresssion_data_
      headers(&thisline, &SIF);
      indices = sid_stp_indices.2.clone();
      SID = sid_stp_indices.0.clone();
      STP = sid_stp_indices.1.clone();
      gene_expression_headers = false;
      continue 'linereading;
    };
  // remaining code...
}
```

In the code in Listing 7-9, the indices are created when the individual identifiers start with "GSM." Once the identifiers are there, the mapping to the description is understood from SIF and then the values STD, STP, and indices are returned to be used later. Once this header line is parsed, we do not want this line to be evaluated anymore and hence we continue the loop after assigning gene_expression_headers to be false so that this block is not executed in anymore of the consecutive line reads.

Now that all the header information has been parsed, we run through the gene data that runs from line 196 to the end in the soft file. This comes under the unique condition of being in the dataset table and not being the gene_expression_headers. Hence we can resort to the following code shown in Listing 7-10.

Listing 7-10. chapter7/statistics/src/main.rs

```
fn process_file(filename: &Path) {
  // previous code ...

  if within_dataset_table && !gene_expression_headers {
    if thisline.starts_with("!dataset_table_end") {
      break 'linereading;
    }
    let (gene_expression_measures, gene_identifiers) = process_
    gene_expresssion_data(&thisline, &indices);
    gene_expression_measures_vec.extend(gene_expression_
    measures);
    gene_identifiers_vec.push(gene_identifiers);
  }

  // remaining code ...
```

process_gene_expresssion_data is the function that will handle this parsing of the lines and putting it in gene_expression_measures_vec and gene_identifiers_vec. The gene_expression_measures_vec will have all the vectors for the different headers and the gene_identifiers_vec will have the concatenated id references and identifiers.

Once the gene_expression_measures_vec has been created, we can convert them to ndaray matrices. In Listing 7-11, we will use the different Array constructs that are provided by ndarray. Array is for an ndarray of arbitrary dimensions, while Array2 and Array1 give us arrays of dimensions 2 and 1 respectively. The function stack can be done to

285

reshape the ndarray to new dimensions, where the Axis function provides
the Axis index.

Listing 7-11. chapter7/statistics/src/main.rs

```
use ndarray;
use ndarray::{Array, Array2, Array1, Axis, stack};

fn convert_to_log_scale(X: &Array2<f64>) -> Array2<f64> {
  let two = 2.0f64;
  let two_log = two.ln();
  X.mapv(|x| x.ln()/two_log)
}

fn process_file(filename: &Path) {
  // previous code ...

  let gene_expression_measures_matrix = Array::from_shape_vec(
    (22283, 127), gene_expression_measures_vec).unwrap();
  let gene_expression_measures_matrix = convert_to_log_scale(
    &gene_expression_measures_matrix);

  // remaining code ...
```

From this matrix, we would need to separate out the UC and Crohn's
disease sample indices. See Listing 7-12.

Listing 7-12. chapter7/statistics/src/main.rs

```
fn filter_specific_samples(
    STP: &Vec<String>,
    group_type: &str) -> Vec<usize> {
  STP.iter().enumerate()
    .filter(|&(_, x)| x == group_type)
    .map(|(i, _)| i).collect()
}
```

```
fn different_samples(
    STP: &Vec<String>) -> (Vec<usize>, Vec<usize>) {
  let UC = filter_specific_samples(&STP, "ulcerative colitis");
  let CD = filter_specific_samples(&STP, "Crohn's disease");
  (UC, CD)
}

let (UC, CD) = different_samples(&STP);
```

Now that we have the data, we can calculate the mean and variance of each group, which will be used to calculated the Z-scores to summarize the evidence of differential expression. To construct the Z-scores, we will need to construct some functions. The first function is to filter out the columns from gene_expression_measures_matrix if we pass the samples indices. The idea is that we should be able to pass the relevant UC and CD indices and get the submatrices. See Listing 7-13.

Listing 7-13. chapter7/statistics/src/main.rs

```
fn filter_out_relevant_columns(samples: &Vec<usize>,
    gene_expression_measures_matrix: &Array2<f64>) ->
    Array2<f64> {
  let shape1 = samples.len();
  let shape0 = gene_expression_measures_matrix.shape()[0];
  let mut cols = Vec::new();
  for &msamples_columns in samples {
    let col = gene_expression_measures_matrix.column(
      msamples_columns);
    cols.push(col);
  }
  let Msamples = stack(Axis(0), &cols[..]).unwrap();
  let Msamples = Array::from_iter(Msamples.iter());
```

```
    let Msamples = Msamples.into_shape((shape0, shape1)).
    unwrap();
    Msamples.mapv(|&x| x)
}
```

Now that we have the submatrices we can either construct the mean of the samples or the variance of the samples, which is a simple function in ndarray (Listing 7-14).

Listing 7-14. chapter7/statistics/src/main.rs

```
fn mean_of_samples(samples: &Vec<usize>,
    gene_expression_measures_matrix: &Array2<f64>) ->
    Array1<f64> {
    let Msamples = filter_out_relevant_columns(
      samples, gene_expression_measures_matrix);
    Msamples.mean_axis(Axis(1))
}

fn variance_of_samples(
    samples: &Vec<usize>,
    gene_expression_measures_matrix: &Array2<f64>)
    -> Array1<f64> {
    let Msamples = filter_out_relevant_columns(samples, gene_
    expression_measures_matrix);
    Msamples.var_axis(Axis(1), 1.)
}
```

Thus, we are finally able to compute the Z-scores for the two samples together (Listing 7-15).

Listing 7-15. chapter7/statistics/src/main.rs

```rust
fn generate_zscores(UC: &Vec<usize>,
                    CD: &Vec<usize>,
                    gene_expression_measures_matrix:
                    &Array2<f64>)
                    -> Array1<f64> {
    let MUC = mean_of_samples(&UC, &gene_expression_measures_
              matrix);
    let MCD = mean_of_samples(&CD, &gene_expression_measures_
              matrix);
    let VUC = variance_of_samples(&UC, &gene_expression_
              measures_matrix);
    let VCD = variance_of_samples(&CD, &gene_expression_
              measures_matrix);
    let nUC = UC.len();
    let nCD = CD.len();
    let z_scores_num = MUC - MCD;
    let z_scores_den = (VUC/nUC as f64 + VCD/nCD as f64).
                       mapv(f64::sqrt);
    let z_scores = z_scores_num / z_scores_den;
                   z_scores
}

fn process_file(filename: &Path) {
  // previous code ...

  let z_scores = generate_zscores(&UC, &CD, &gene_expression_
                 measures_matrix);
  let z_scores_mean = z_scores.sum() / z_scores.len() as f64;
  let z_scores_std = z_scores.std_axis(Axis(0), 1.);
  println!("z scores mean {:?}", z_scores_mean);
  println!("z scores mean {:?}", z_scores_std);
}
```

If a gene is not differentially expressed, it has the same expected values in the two groups of samples. In this case the Z-scores will be standardized or will have a zero mean and unit standard deviation. Printing out z_scores_mean and z_scores_std we find that the values are 0.042918212228268235 and 3.5369539066322027 respectively. Notice that since the vectors are in ndarray format, we can easily compute the stats using the ndarray api.

Since the standard deviation is much greater than 1, there appear to be multiple genes for which the mean expression levels in the UC and Crohn's disease samples differ. Further, since the mean Z-score is positive, it appears that the dominant pattern is for genes to be expressed more in the UC compared to the Crohn's disease samples.

Similarly we can find the p-values. The exact results are inconsequential for this section, the aim of which is to show that once values are converted to ndarray matrices in Rust, it is fairly trivial to perform statistical analysis on them.

7.2 Writing High Performance Code

A lot of the machine learning code that has the ability to use GPU is through Rust bindings to libraries that are essentially created in C/C++. To be able to use the GPU or write high performance code, we can use crates such as blas-src or lapack-src, which provide access to low-level mathematical operations.

In the majority of the cases the high-level crates and libraries that we have discussed in the previous chapters would be enough, but sometimes we would need to go low-level and perform direct vector operations. The Single Instruction Multiple Data operation library, or more popularly known as SIMD, is a popular library for performing vector operations, and the faster[1] crate gives us good abstractions for running SIMD instructions in Rust.

---

[1]https://github.com/AdamNiederer/faster.

SIMD and faster are great if we are trying to get parallel instructions being run on a vector and to return the result. The transformation runs on each element of the vector. In the code shown in Listing 7-16, we will take a vector and return the cube of all the elements in the vector.

Before running the installation code, you should probably have blas and lapack libraries installed in the system. Installing in an ubuntu system might need a command such as that shown in this code.

Listing 7-16. Installing libblas

```
$ sudo apt-get install libblas-dev liblapack-dev
$ sudo apt-get install libopenblas-dev
$ sudo apt-get install gfortran
$
```

We can now create a Rust project to explore how to write high performance code and name it high performance computing (Listing 7-17).

Listing 7-17. Create Rust project for high performance computing

```
$ cd chapter7
$ cargo new high-performance-computing --bin && cd high-
  performance-computing
```

We should now have the main.rs file and the Cargo.toml file. Since we will be using lapack, openblas, and SIMD dependencies, we would need to add the crates that enable us to use those dependencies. See Listing 7-18.

Listing 7-18. chapter7/high-performance-computing/Cargo.toml

```
[package]
name = "high-performance-computing"
version = "0.1.0"
edition = "2018"
```

```
[dependencies]
faster = "0.5.0"
rblas = "0.0.13"
```

In Listing 7-18, the dependencies list, the crate faster is the Rust code for writing SIMD code in Rust, rblas crate gives us access to the open BLAS and OpenBLAS packages.

We should now be able to write the code and we will start with using the faster package and writing code that uses the SIMD software (Listing 7-19).

Listing 7-19. chapter7/high-performance-computing/src/main.rs

```rust
use faster::*;

fn main() {
    let my_vector: Vec<f32> = (0..10).map(
        |v| v as f32).collect();
    let power_of_3 = (&my_vector[..]).simd_iter(f32s(0.0))
        .simd_map(|v| v * v * v)
        .scalar_collect();
    println!("{:?}", power_of_3);
}
```

Similarly when we are trying to perform a reduction on the vector, we can use the simd_reduce method. For example, in the code in Listing 7-20, we will try to find the sum of the elements in a vector.

Listing 7-20. chapter7/high-performance-computing/src/main.rs

```rust
fn main() {
    // previous code ...

    let reduced = (&power_of_3[..])
        .simd_iter(f32s(0.0))
        .simd_reduce(f32s(0.0), |a, v| a + v ).sum();
}
```

Another common operation that is performed on vectors is the dot product of two vectors, and in that case we will use the rblas crate, included earlier, which is a wrapper over blas and openblas libraries. BLAS, which stands for Basic Linear Algebra Specification, is a set of low-level routines for performing common linear operations. The rblas crate has the Dot product implemented over blas [1].

In Listing 7-21, we take the dot product of two vectors initialized.

Listing 7-21. chapter7/high-performance-computing/src/main.rs

```
use rblas::Dot;

fn main() {
    let x = vec![1.0, -2.0, 3.0, 4.0];
    let y = [1.0, 1.0, 1.0, 1.0, 7.0];

    let d = Dot::dot(&x, &y[..x.len()]);
    println!("dot product {:?}", d);
}
```

Running the code in Listing 7-21, we should get the output in Listing 7-22.

Listing 7-22. high-performance-computing output

```
$ cargo run
    Finished dev [unoptimized + debuginfo] target(s) in 0.01s
    Running `target/debug/high-performance-computing`
[3.0, 3.0, ..., 3.0, 3.0]
[0.0, 1.0, 8.0, 27.0, 64.0, 125.0, 216.0, 343.0, 512.0, 729.0]
2025.0
dot product 6.0
```

Thus using these wrappers over high performance libraries, we should be able to write high performance code in our Rust applications.

7.3 Recommender Systems

Recommender systems are one of the most successful and widespread applications of machine learning technologies in business. Machine learning algorithms in recommender systems are typically classified into two categories – content-based filtering and collaborative filtering although modern architectures mainly employ a combination of both. Content-based methods are based on similarity of item attributes and collaborative methods calculate similarity from interactions.

In Rust, we are able to create recommendation engines mostly due to the great sbr-rs crate.[2] In this crate there are two models that are implemented. This crate will need some additional OS-level dependencies installed. Ensure that you have libssl installed in your system. In an ubuntu system, running the following command in Listing 7-23 should be fine.

Listing 7-23. Install libssl ubuntu

```
$ sudo apt install libssl-dev
$
```

The models that we can create using the package sbr are these:

- LSTM: an LSTM network is used to model the sequence of a user's interaction to predict their next action;

- EWWA: This model uses an exponentially weighted average of past actions to predict future interactions.

To start the project, we can create a project using cargo as shown in Listing 7-24.

---

[2]https://github.com/maciejkula/sbr-rs/.

Listing 7-24. New rust goodbooks package

```
$ cargo new --bin goodbooks-recommender
$
```

This will set up the project with the src/main.rs file and the Cargo.toml files as we have seen in the previous chapters. Add the dependencies in the cargo file. Some of the dependencies we have already seen such as reqwest to serve as a web client. Failure is an excellent crate for better handling of errors. The crates serde, serde_json, and serde_json are there to be able to serialize and deserialize the data. We will need the csv crate for working with csv files. The crate rand will help us in randomizing the data and structopt crate to create good command-line interfaces. The most important crate as per this section is the sbr crate, which has the recommendation modules implemented that we will call here in Listing 7-25.

Listing 7-25. chapter7/goodbooks-recommender/Cargo.toml

```
[package]
name = "goodbooks-recommender"
version = "0.1.0"
authors = ["author names"]
edition = "2018"

[dependencies]
reqwest = "0.9.17"
failure = "0.1.5"
serde = "1"
serde_derive = "1"
serde_json = "1"
csv = "1"
sbr = "0.4.0"
rand = "0.6.5"
structopt = "0.2.15"
```

7.3.1 Command Line

We will now start with the creation of the command line. For readers who are familiar with git, you might have noticed that git has first-level commands such as git clone or git pull, and then depending on the first command, we will have a second level of commands such as origin master like a subcommand for pull, which enables us to write git pull origin master. Along similar lines we will try to build a command-line parser (Listing 7-26).

Listing 7-26. goodbooks-recommender/src/main.rs

```rust
use structopt::StructOpt;

#[derive(Debug, StructOpt)]
#[structopt(name = "goodbooks-recommender", about = "Books
Recommendation")]
enum Opt {
  #[structopt(name = "fit")]
  /// Will fit the model.
  Fit,
  #[structopt(name = "predict")]
  /// Will predict the model based on the string after this.
  Please run
  /// this only after fit has been run and the model has been
  saved.
  Predict(BookName),
}

// Subcommand can also be externalized by using a 1-uple enum
variant
#[derive(Debug, StructOpt)]
```

```
struct BookName {
  #[structopt(short = "t", long = "text")]
  /// Write the text for the book that you want to predict
  /// Multiple books can be passed in a comma separated manner
  text: String,
}

fn main() {
  let opt = Opt::from_args();
}
```

In the code in Listing 7-26, the goodbooks-recommender takes in a first command, which is essentially an enum of either Fit or Predict. Fit is fine by itself but Predict takes in a separate subcommand to it which is a text. The help function also gives documentation on this (Listing 7-27).

Listing 7-27. Package run help

```
$ cargo run -- --help
    Finished dev [unoptimized + debuginfo] target(s) in 0.23s
     Running `target/debug/goodbooks-recommender --help`
goodbooks-recommender 0.1.0
Books Recommendation

USAGE:
    goodbooks-recommender <SUBCOMMAND>

FLAGS:
    -h, --help       Prints help information
    -V, --version    Prints version information
```

```
SUBCOMMANDS:
    fit        Will fit the model.
    help       Prints this message or the help of the given
               subcommand(s)
    predict    Will predict the model based on the string after
               this. Please run this only after fit has been run
               and the model has been saved.
```

We are also able to see the help text for predict as shown in Listing 7-28.

Listing 7-28. Package run predict help

```
$ cargo run -- predict --help
    Finished dev [unoptimized + debuginfo] target(s) in 0.23s
     Running `target/debug/goodbooks-recommender predict
    --help`
goodbooks-recommender-predict 0.1.0
Will predict the model based on the string after this. Please
run this only after fit has been run and the model has
been saved.

USAGE:
    goodbooks-recommender predict --text <text>

FLAGS:
    -h, --help       Prints help information
    -V, --version    Prints version information

OPTIONS:
    -t, --text <text>    Write the text for the book that you
                         want to predict Multiple books can be
                         passed in a comma separated manner
```

7.3.2 Downloading Data

We should now be able to create functions for downloading the data and create saving the files in the destination folders (Listing 7-29).

Listing 7-29. chapter7/goodbooks-recommender/src/main.rs

```
use std::fs::File;
use std::io::BufWriter;
use std::path::Path;

fn download(url: &str, destination: &Path)
            -> Result<(), failure::Error> {

  if destination.exists() {
    return Ok(()) // do not download multiple times
  }

  let mut writer = BufWriter::new(file);
  let mut response = reqwest::get(url)?;
  response.copy_to(&mut writer)?;

  Ok(())
}
```

In the above download function shown in Listing 7-29, we take the url as a str and destination as a path. If the path exists, then nothing needs to be done. We will leave to the caller of this function to ensure that the url and destination combinations are correct. Then we will get the response from the url and write it to the path using BufWriter module. This should download any url that we pass to the function.

Using the above function, we can download the good books dataset from this github repo https://github.com/zygmuntz/goodbooks-10k/. This dataset contains six million ratings for ten thousand most of the popular books. The other kinds of information that are provided in the

datasets are such as isbn values and authors and so on, but we will not
be concerned with them for the purpose of this example. We will be
downloading the ratings.csv and the books.csv as seen in Listing 7-30.

Listing 7-30. chapter7/goodbooks-recommender/src/main.rs

```
/// download ratings and metadata both.
fn download_data(ratings_path: &path,
                 books_path: &path) {
  let ratings_url = "https://github.com/zygmuntz/\
                     goodbooks-10k/raw/master/ratings.csv";
  let books_url = "https://github.com/zygmuntz/\
                   goodbooks-10k/raw/master/books.csv";

  download(&ratings_url, ratings_path)
    .expect("could not download ratings");
  download(&books_url, books_path)
    .expect("could not download metadata");
}
```

7.3.3 Data

Now we can run this function and we will see the files downloaded in the
root folder. Take a look at these two csv's. The columns in books.csv file
are book_id, goodreads_book_id, work_id, books_count, isbn, isbn13,
authors, original_publication_year, language_code, title, language_
code, average_rating, ratings_count, work_ratings_count, work_text_
reviews_count, ratings_1, ratings_2, ratings_3, ratings_4, ratings_5,
image_url, and small_image_url, thus having 23 columns. Similarly, the
ratings file has user_id, book_id, and rating.

From this ratings file we will create the mapping between user_id and
book_id and from the book file we will create the mapping between the
book_id and the title. This can be encoded in a struct. We are only taking

a small number of features for the sake of simplicity, but you should try it out with more features. See Listing 7-31.

Listing 7-31. chapter7/goodbooks-recommender/src/main.rs

```
#[derive(Debug, Serialize, Deserialize)]
struct UserReadsBook {
  user_id: usize,
  book_id: usize,
}

#[derive(Debug, Deserialize, Serialize)]
struct Book {
  book_id: usize,
  title: String
}
```

We can now write two functions, one for deserializing the ratings and one for deserializing the books (Listing 7-32).

Listing 7-32. chapter7/goodbooks-recommender/src/main.rs

```
use csv;

fn deserialize_ratings(path: &Path)
        -> Result<Vec<UserReadsBook>, failure::Error> {
  let mut reader = csv::Reader::from_path(path)?;
  let entries = reader.deserialize()
    .collect::<Result<Vec<_>, _>>()?;

  Ok(entries)
}

fn deserialize_books(path: &Path)
    -> Result<(HashMap<usize, String>,
               HashMap<String, usize>), failure::Error> {
```

```
let mut reader = csv::Reader::from_path(path)?;
let entries: Vec<Book> = reader.deserialize::<Book>()
  .collect::<Result<Vec<_>, _>>()?;

let id_to_title: HashMap<usize, String> = entries
  .iter()
  .map(|book| (book.book_id, book.title.clone()))
  .collect();
let title_to_id: HashMap<String, usize> = entries
  .iter()
  .map(|book| (book.title.clone(), book.book_id))
  .collect();

Ok((id_to_title, title_to_id))
}
```

In both Listings 7-31 and 7-32, we will read the files using csv and then collect the vectors. In the deserialize_ratings instead of directly returning the vectors, we will return the result of the vectors given by Result<Vec<UserReadsBook>. In the deserialize_books function, additionally we create id_to_title and title_to_id mappings so that we are able to get one from the other through a simple lookup on the relevant mapping.

7.3.4 Model Building

The data part already is already done, so now we can compose functions for building the models. As stated earlier, the sbr implements two models that we can use an LSTM-based model and a moving average-based model (EWMA). We will go ahead and use the EWMA model. The EWMA model is simpler in terms of computational weight and we will use this here. An exponentially weighted moving average is a type of control used to monitor small shifts in the process mean. It weights observations in geometrically decreasing order so that the most recent observations carry the most weight and the older observations contribute very little to the model. In many cases this is enough.

First we write a function that sets up the hyperparameters of the model (Listing 7-33).

Listing 7-33. chapter7/goodbooks-recommender/src/main.rs

```rust
use sbr::models::ewma::{Hyperparameters, ImplicitEWMAModel};
use sbr::models::{Loss, Optimizer};

fn build_model(num_items: usize) -> ImplicitEWMAModel {
  let hyperparameters = Hyperparameters::new(num_items, 128)
    .embedding_dim(32)
    .learning_rate(0.16)
    .l2_penalty(0.0004)
    .loss(Loss::WARP)
    .optimizer(Optimizer::Adagrad)
    .num_epochs(10)
    .num_threads(1);

  hyperparameters.build()
}
```

For the model to work, we need the data converted to sbr interactions objects. Interactions take in a number of users and a number of items and the timestamp. Since the ids are ordinal, we can take the greatest ids to be the number of users and items and have the id as the timestamp assuming that the data is ordered chronologically. In other situations, we will need to take care of these in a different manner (Listing 7-34)

Listing 7-34. chapter7/goodbooks-recommender/src/main.rs

```
fn build_interactions(data: &[UserReadsBook]) -> Interactions {
  let num_users = data
    .iter()
    .map(|x| x.user_id)
    .max()
    .unwrap() + 1;
  let num_items = data
    .iter()
    .map(|x| x.book_id)
    .max()
    .unwrap() + 1;
  let mut interactions = Interactions::new(num_users,
                                           num_items);
  for (idx, datum) in data.iter().enumerate() {
    interactions.push(
      Interaction::new(datum.user_id,
                       datum.book_id,
                       idx)
    );
  }
  interactions
}
```

Now that the model has been built, we will need to train on the data. We will now go ahead and create a fit function. As with other training done before, the data needs to be split into test and train so that the fitness of the model can be determined and we can understand that the model is actually learning. See Listing 7-35.

Listing 7-35. chapter7/goodbooks-recommender/src/main.rs

```
use sbr::evaluation::mrr_score;

fn fit(model: &mut ImplicitEWMAModel,
       data: &Interactions)
       -> Result<f32, failure::Error> {

  let (train, test) = user_based_split(
        data, &mut rng, 0.2);
  model.fit(&train.to_compressed())?;

  let mrr = mrr_score(model, &test.to_compressed())?;
  Ok(mrr)
}
```

In the fit function in Listing 7-35, the result is the mrr score. The MRR score or the mean reciprocal score is the score when the validity of the single highest-ranking item is measured. Unfortunately, the other popular means of scoring, which is the mean average precision, is not implemented in sbr and would need to be implemented by the user.

Once the model training is done, we will need to serialize the model, so that we can save the model in a file. We can use the serde library to do this (Listing 7-36).

Listing 7-36. chapter7/goodbooks-recommender/src/main.rs

```
fn serialize_model(model: &ImplicitEWMAModel,
                   path: &Path) -> Result<(), failure::Error> {
  let file = File::create(path)?;
  let mut writer = BufWriter::new(file);

  Ok(serde_json::to_writer(&mut writer, model)?)
}
```

We will now need a function to execute the functions we just mentioned, one after the other. See Listing 7-37.

Listing 7-37. chapter7/goodbooks-recommender/src/main.rs

```rust
fn main_build() {
    let ratings_path = Path::new("ratings.csv");
    let books_path = Path::new("books.csv");
    let model_path = Path::new("model.json");

    // Exit early if we already have a model.
    if model_path.exists() {
        println!("Model already fitted.");
        return ();
    }

    download_data(ratings_path, books_path);

    let ratings = deserialize_ratings(ratings_path).unwrap();
    let (id_to_title,
        title_to_id) = deserialize_books(books_path).unwrap();

    println!("Deserialized {} ratings.", ratings.len());
    println!("Deserialized {} books.", id_to_title.len());

    let interactions = build_interactions(&ratings);
    let mut model = build_model(interactions.num_items());

    println!("Fitting...");
    let mrr = fit(&mut model, &interactions)
        .expect("Unable to fit model.");
    println!("Fit model with MRR of {:.2}", mrr);

    serialize_model(&model, &model_path)
        .expect("Unable to serialize model.");
}
```

We should now plug in this main_build function in the main function (Listing 7-38).

Listing 7-38. chapter7/goodbooks-recommender/src/main.rs

```rust
fn main() {
  let opt = Opt::from_args();
  match opt {
    Opt::Fit => main_build(),
    _ => {
      unimplemented!();
    },
  }
}
```

Running it now with cargo run - fit should save the model in the model.json file (Listing 7-39).

Listing 7-39. Check model.json in directory

```
$ ls
books.csv  Cargo.lock  Cargo.toml  fit  model.json  ratings.
csv  src  target
$
```

7.3.5 Model Prediction

To be able to do predictions, we need to be able to do two things. First is model deserialization. After model deserialization, the model will be in ImplicitEWMAModel struct (Listing 7-40).

Listing 7-40. chapter7/goodbooks-recommender/src/main.rs

```
use std::io::BufReader;

fn deserialize_model() -> Result<ImplicitEWMAModel,
                                  failure::Error> {
  let file = File::open("model.json")?;
  let reader = BufReader::new(file);

  let model = serde_json::from_reader(reader)?;

  Ok(model)
}
```

After model deserialization is done, we will use the model to make predictions. So the predict function needs to be passed to the model for inference. Also the target tokens need to be passed (Listing 7-41).

Listing 7-41. chapter7/goodbooks-recommender/src/main.rs

```
fn predict(input_titles: &[String],
           model: &ImplicitEWMAModel)
           -> Result<Vec<String>, failure::Error> {
  let (id_to_title,
       title_to_id) = deserialize_books(
       &Path::new("books.csv")
  ).unwrap();

  // Let's first check if the inputs are valid.
  for title in input_titles {
    if !title_to_id.contains_key(title) {
      println!("No such title, ignoring: {}", title);
    }
  }
  // rem code..
}
```

For this we will need to create the user representation and the possible indices that can be predicted.

Since the model is trained on the title ids, we will provide the ids as input for prediction. We will also need the possible ids that can be predicted. In this case the ids are ordered and hence we can just take from 0 to the length. In other cases, we would have needed to collect the actual vector. See Listing 7-42.

Listing 7-42. chapter7/goodbooks-recommender/src/main.rs

```
fn predict(input_titles: &[String],
           model: &ImplicitEWMAModel)
           -> Result<Vec<String>, failure::Error> {
  // prev code..

  // Map the titles to indices.
  let input_indices: Vec<_> = input_titles
    .iter()
    .filter_map(|title| title_to_id.get(title))
    .cloned()
    .collect();
  let indices_to_score: Vec<usize> =
    (0..id_to_title.len()).collect();

  // rem code..
}
```

Based on the input_indices vector, we can get the user representation that will be passed to the model for predictions (Listing 7-43).

Listing 7-43. chapter7/goodbooks-recommender/src/main.rs

```rust
fn predict(input_titles: &[String],
           model: &ImplicitEWMAModel)
           -> Result<Vec<String>, failure::Error> {
  // prev code ..

  // Get the user representation.
  let user = model.user_representation(&input_indices)?;
  // Get the actual predictions.
  let predictions = model.predict(&user, &indices_to_score)?;

  // rem code ..
```

Once the predictions are generated, we will need to sort based on the
score. Here we will show only the top 10 results (Listing 7-44).

Listing 7-44. chapter7/goodbooks-recommender/src/main.rs

```rust
fn predict(input_titles: &[String],
           model: &ImplicitEWMAModel)
           -> Result<Vec<String>, failure::Error> {
    // prev code ...

    let mut predictions: Vec<_>
        = indices_to_score.iter()
          .zip(predictions)
          .map(|(idx, score)| (idx, score))
          .collect();

    predictions
      .sort_by(
        |(_, score_a), (_, score_b)|
          score_b.partial_cmp(score_a)
          .unwrap());
```

```
    Ok((&predictions[..10])
        .iter()
        .map(|(idx, _)| id_to_title.get(idx).unwrap())
        .cloned()
        .collect())
}
```

Plugging it in the main method should yield the results we need (Listing 7-45).

Listing 7-45. chapter7/goodbooks-recommender/src/main.rs

```
fn main() {
  let opt = Opt::from_args();
  match opt {
    Opt::Fit => main_build(),
    Predict(book) => {
      let model = deserialize_model()
        .expect("Unable to deserialize model.");
      let tokens: Vec<String> = book.text.split(",").map(
        |s| s.to_string()).collect();
      let predictions = predict(&tokens, &model)
        .expect("Unable to get predictions");
      println!("Predictions:");
      for prediction in predictions {
        println!("{}", prediction);
      }
    },
  }
}
```

Using the command in Listing 7-46, we can predict user input titles.

Listing 7-46. chapter7/goodbooks-recommender/src/main.rs

```
$ cargo run -- predict --text "The Alchemist"
    Finished dev [unoptimized + debuginfo] target(s) in 0.22s
     Running `target/debug/goodbooks-recommender predict --text
     'The Alchemist'`
Predictions:
The Alchemist
The Kite Runner
One Hundred Years of Solitude
The Da Vinci Code (Robert Langdon, #2)
A Thousand Splendid Suns
Life of Pi
Eat, Pray, Love
Memoirs of a Geisha
Angels & Demons  (Robert Langdon, #1)
The Five People You Meet in Heaven
$
```

7.4 Conclusion

This chapter introduced you to different interesting domains in high
performance computing with Rust that are generally nontraditional
applications of machine learning. The chapter starts with statistical
analysis and how statistical analysis becomes trivial once we are able to
convert datasets to ndarray matrices. In the next section, we used wrappers
of SIMD and BLAS to perform high performance and parallel computation
on vectors. Finally, we built a recommendation system using a popular
recommendation crate in Rust.

In the last chapter of the book, you will learn how we can create and
deploy Rust ML applications for production using cloud as well as other
ways of deploying Rust machine learning applications and models.

7.5 Bibliography

[1] Various. *Recommender system*. [Online; accessed 23-May-2019]. 2019. URL: `https://en.wikipedia.org/wiki/Recommender_system`.

[2] Pavel Kordík. *Machine Learning for Recommender systems — Part 1 (algorithms, evaluation and cold start)*. Ed. by Towards Data Science. [Online; accessed 23-May-2019]. 2018. URL: `https://medium.com/recombee-blog/machine-learning-for-recommender-systems-part-1-algorithmsevaluation-and-cold-start-6f696683d0ed`.

[3] Maciej Kula. *Recommending books (with Rust)*.

[4] Jason Brownlee. *A Gentle Introduction to Vectors for Machine Learning*. Ed. by Machine Learning Mastery. [Online; accessed 1-June-2019]. 2018. URL: `https://machinelearningmastery.com/gentle-introductionvectors-machine-learning/`.

CHAPTER 8

Using Rust Applications

Throughout this book, we have looked into data transformations and creating machine learning models in Rust. Once we know the machine learning workflow that we are going ahead with, we need a way to integrate the workflow in our overall architecture. Since Rust is a fairly new language, it is highly probable that overall architecture in your project is probably created in a mainstream language. In this chapter we will look at how to integrate Rust code into our overall production architecture. We will start with integrating Rust code in Python applications and then we will move on to integrating Rust code in Java applications. We shall finally take a look at creating serverless applications in Rust.

8.1 Rust Plug-n-Play

Since Rust is a relatively new application, chances are quite high that you are working in a legacy application with a couple of million lines of code, written in a popular language such as Python or Java. In that case, using a Rust ML application would mean only using a small part of the existing application. The fact that Rust does not have a runtime makes it very easy

© Joydeep Bhattacharjee 2020
J. Bhattacharjee, *Practical Machine Learning with Rust*,
https://doi.org/10.1007/978-1-4842-5121-8_8

to call Rust code in other languages as long as these "other" languages have a way of calling shared libraries. Generally, all mainstream languages have such capabilities. In this next sections, we will take two examples, one in which we will call Rust functions in Python and one in Java.

8.1.1 Python

In this section, we will run call Rust code using PyO3.[1] PyO3 provides Rust bindings for Python. In this code we will take the `crfsuite-model` that was created in Chapter 5 and try to call the code from Python.

First let's recap a little bit. In the crfsuite code, we have a struct `NER` to read `lemma`, `next_lemma`, `word`, and `tag` from the dataset. The data is then passed through different functions such as `split_test_train` and `create_xseq_yseq` to convert to the correct sequence of vectors. We can then pass the data to `crfmodel_training`, which will perform the training. This training also creates a model file. Once trained, we can use `model_prediction` to perform the prediction and use `check_accuracy` function to check the accuracy.

To be able to use these functionalities, we will need to create a public api and wrap them in some Python callable code. So let's first look at the dependencies and the overall project structure that we will need. We will need the data reading and organization crates that we have seen before `csv`, `serde`, `serde-derive`, and `rand`. We will also need the `crfsuite` for being able to perform machine learning and run the previous code. We will also need the latest code for pyo3. Finally, we will need to specify the library name and crate type. If we don't use dylib, then it will create rlib binaries. The format rlib is a Rust-specific static library format that includes metadata such as serialized typechecked AST-s for generics. Hence they are not suitable for external consumption. Also, don't use

---

[1]https://github.com/PyO3/pyo3.

staticlib, since this fails to link as well. Most other languages understand and produce .so/.dylib formats only. Thus, we will have the Cargo.toml similar to Listing 8-1.

Listing 8-1. chapter8/crfsuite-model/Cargo.toml

```
[package]
name = "crfsuite-model"
version = "0.2.0"
edition = "2018"

[dependencies]
csv = "1.0.7"
serde = "1"
serde_derive = "1"
rand = "0.6.5"
crfsuite = "0.2.6"
pyo3 = { git = "https://github.com/PyO3/pyo3.git",
    // Take the latest revision from github
    rev = "99fdafbb880c181f4bce16bbbac03888b3cf85c8",
    features = ["extension-module"]}

[lib]
name = "crfsuite_model"
crate-type = ["cdylib"]
```

Now we will need to decide on a public api for the code. This is similar to the famous scikit-learn style of creating a model class and defining fit and predict methods on the code. The class can take the model file name as input and train and predict can take the path of the training and predict files as the data to train and predict (Listing 8-2).

Listing 8-2. chapter8/crfsuite-model/src/lib.rs

```rust
use pyo3::prelude::*;
use std::fs;
use std::path::PathBuf;

#[pyclass(module = "crfsuite_model")]
pub struct CRFSuiteModel {
  model_name: String,
}

#[pymethods]
impl CRFSuiteModel {
  #[new]
  fn new(obj: &PyRawObject, path: String) {
    obj.init(CRFSuiteModel {
      model_name: path,
    });
  }

  fn fit(&self, py: Python<'_>, path: String) -> PyResult<String> {
    // code for training ...
    Ok("model fit done".to_string())
  }

  fn predict(&self,
             predict_filename: String)
             -> PyResult<Vec<String>> {
    // code for predict ...
    Ok(preds)
  }
}
```

Now that we have the overall function signature, transfer the main method that we had in version 0.1 to the fit method (Listing 8-3).

Listing 8-3.   chapter8/crfsuite-model/src/lib.rs

```rust
fn fit(&self, py: Python<'_>, path: String) -> PyResult<String> {
  let data_file = PathBuf::from(
        &path[..]);
  let data = get_data(
        &data_file).unwrap();
  let (test_data, train_data) = split_test_train(
        &data, 0.2);
  let (xseq_train, yseq_train) = create_xseq_yseq(
        &train_data);
  let (xseq_test, yseq_test) = create_xseq_yseq(
        &test_data);
  crfmodel_training(
        xseq_train, yseq_train, self.model_name.as_ref())
        .unwrap();
  let preds = model_prediction(
        xseq_test, self.model_name.as_ref())
        .unwrap();
  check_accuracy(&preds, &yseq_test);
  Ok("model fit done".to_string())
}
```

The description of the functions split_test_train, create_xseq_yseq, crfmodel_training, model_prediction, and check_accuracy was described in Chapter 5.

To be able to create the predict method, we will need to create some additional code as the predict code cannot have the functions create_xseq_yseq because the test file will not have labels for the x sequences.

Once the x_seq is created, we can pass the sequences to model_prediction to get the predictions.

We will need to create a new struct that will not have labels. We can then have similar functions such as get_data for reading the data as per the predict function and create_xseq_yseq for creating the x_seq's without the labels (Listing 8-4).

Listing 8-4. chapter8/crfsuite-model/src/lib.rs

```rust
#[derive(Debug, Deserialize, Clone)]
pub struct NER_Only_X {
  lemma: String,
  #[serde(rename = "next-lemma")]
  next_lemma: String,
  word: String,
}

fn get_data_no_y(path: &PathBuf) -> Result<Vec<NER_Only_X>,
Box<dyn Error>> {
  let csvfile = fs::File::open(path)?;
  let mut rdr = csv::Reader::from_reader(csvfile);
  let mut data = Vec::new();
  for result in rdr.deserialize() {
    let r: NER_Only_X = result?;
    data.push(r);
  }
  Ok(data)
}

fn create_xseq_for_predict(data: &[NER_Only_X])
        -> Vec<Vec<Attribute>> {
  let mut xseq = vec![];
```

```
for item in data {
  let seq = vec![Attribute::new(item.lemma.clone(), 1.0),
  Attribute::new(item.next_lemma.clone(), 0.5)];
  // higher weightage for the mainword.
  xseq.push(seq);
}
xseq
}
```

We should now be able to stitch these functions in the predict method (Listing 8-5).

Listing 8-5. chapter8/crfsuite-model/src/lib.rs

```
fn predict(&self, predict_filename: String) ->
PyResult<Vec<String>> {
  let predict_data_file = PathBuf::from(
        predict_filename);
  let data = get_data_no_y(
        &predict_data_file).unwrap();
  let xseq_test = create_xseq_for_predict(
        &data[..]);
  let preds = model_prediction(
        xseq_test, self.model_name.as_ref())
        .unwrap();
  Ok(preds)
}
```

Now we will need to register the applications as a Python module (Listing 8-6).

Listing 8-6. chapter8/crfsuite-model/src/lib.rs

```
#[pymodule]
fn crfsuite_model(_py: Python<'_>, m: &PyModule) ->
PyResult<()> {
  m.add_class::<CRFSuiteModel>()?;
  Ok(())
}
```

Now that the Rust api is done, we will need to call the application from Python and we will need to set up some additional plumbing. We would need a setup file to run the cargo build file, a pyproject.toml, and a MANIFEST.in so that the appropriate files are bundled.

The `setup.py` file is the same as the pyo3 examples.[2] The only differences are in the project file names (Listing 8-7).

Listing 8-7. chapter8/crfsuite-model/setup.py

```
setup(
  name="crfsuite-model",
  version="0.2.0",
  classifiers=[
    "License :: OSI Approved :: MIT License",
    "Development Status :: 3 - Alpha",
    "Intended Audience :: Developers",
    "Programming Language :: Python",
    "Programming Language :: Rust",
    "Operating System :: POSIX",
    "Operating System :: MacOS :: MacOS X",
  ],
  packages=["crfsuite_model"],
```

---

[2]https://github.com/PyO3/pyo3/tree/master/examples/word-count.

```
rust_extensions=[RustExtension("crfsuite_model.crfsuite_
model", "Cargo.toml")],
install_requires=install_requires,
tests_require=tests_require,
setup_requires=setup_requires,
include_package_data=True,
zip_safe=False,
cmdclass={
  'test': PyTest,
  'sdist': CargoModifiedSdist,
},
)
```

Listing 8-8. chapter8/crfsuite-model/MANIFEST.in

```
include pyproject.toml Cargo.toml
recursive-include src *
```

Listing 8-9. chapter8/crfsuite-model/pyproject.toml

```
[build-system]
requires = ["setuptools>=41.0.0", "wheel", "setuptools_
rust>=0.10.2", "toml"]
build-backend = "setuptools.build_meta"
```

We will also need to create a director with the project name and place an __init__.py file there. This would help in calling the built shared object file that can be accessed using a clean interface (Listing 8-10).

Listing 8-10. init

```
from .crfsuite_model import CRFSuiteModel

__all__ = ["CRFSuiteModel",]
```

Since we are organizing the files in a package, we should be able to call the classes as if the classes had been written in pure Python. Next are training (Listing 8-11) and predict code (Listing 8-12) as different files. Note that we are passing the same model name in the code.

Listing 8-11. Example training file

```
# coding: utf-8
from crfsuite_model import CRFSuiteModel
model = CRFSuiteModel("model.crfsuite")
res = model.fit("data/ner.csv")
print(res)
```

Listing 8-12. Example prediction file

```
# coding: utf-8
from crfsuite_model import CRFSuiteModel
model = CRFSuiteModel("model.crfsuite")
res = model.predict("data/ner_predict.csv")
print(res)
```

Now we should be able to run the training and prediction Python code (Listing 8-13).

Listing 8-13. Run training code

```
$ cd chapter8
$ python3 —m venv venv
$ source venv/bin/ activate
( venv ) $ pip install -e .
( venv ) $ python crfsuite_model_training.py
Feature generation
type: CRF1d
```

```
feature . minfreq : 0.000000
feature . possible_states : 0
feature . possible_transitions : 0
0 .... 1 .... 2 .... 3 .... 4 .... 5 .... 6 .... 7 .... 8
.... 9 .... 10
Number of features : 3136
Seconds required : 0.014

Adaptive Regularization of Weights (AROW)
variance : 1.000000
gamma: 1.000000
max_iterations : 100
epsilon : 0.000000

***** Iteration #1 *****
Loss : 1214.000000
Feature norm : 0.526245
Seconds required for this iteration : 0.014

***** Other iterations *****

***** Iteration #100 *****
Loss : 455.980399
Feature norm : 329.397943
Seconds required for this iteration : 0.012

Total seconds required for training : 1.257

Storing the model
Number of active features : 3115 (3136)
Number of active attributes : 2065 (2086)
Number of active labels : 17 (17)
Writing labels
Writing attributes
```

```
Writing feature references for transitions
Writing feature references for attributes
Seconds required : 0.015

accuracy =0.5263158 (160/304 correct)
model fit done
$
```

If we now check the directory, the crfsuite model file named "model. crfsuite" should be created (Listing 8-14).

Listing 8-14. Check for model file

```
$ l s
Cargo . lock      crfsuite_model.egg–info      data
requirements–dev.txt      venv
Cargo . toml      crfsuite_model_prediction.py   model.crfsuite
setup . py
MANIFEST. in      crfsuite_model_training.py     pip–wheel–meta
data src
crfsuite_model    crfsuite_model_training1.py    pyproject.toml
target
$
```

Now that the model is created, we should be able to run the prediction code (Listing 8-15).

Listing 8-15. Run prediction code

```
( vevn ) $ python crfsuite_model_prediction . py
[ '0' , '0' , '0' , '0' , '0' , '0' , 'B–geo' , '0', '0' ]
$
```

Similarly, we can take other examples from this book and create apis that are callable in Python.

8.1.2 Java

Similar to the Python example that we had a look at earlier, we can integrate Rust functions and call them in a Java class. The concepts are similar where we will need to expose the Rust functions in a C interface, and in Java we will define native methods that are referenced to those C interfaces. This is done using the jni crate. Let us take the xgboost example in Chapter 2 and augment it by making it a Java application.

Similar to what we have seen in the Python section, any ffi project will have some code in both of the languages. We will see the least amount of code that is required to act as the bridge between the Java side and the Rust side. Java requires all native methods to adhere to the Java Native Interface or the JNI standard. So we will need to define the function signature from Java, and then we can write the Rust code that will adhere to it. The steps are the following:

- Write the functionality;

- Write the Java class;

- Write the Rust interface that wraps the Rust functionality and provides access to those functionalities in Java.

Rust functionality To implement the Rust functions, we will copy and paste the previous code in xgboost. Notice that the previous code was a binary executable. Hence we will need to convert the package to a library to be able to call the functions in Rust (Listing 8-16).

Listing 8-16. Copy xgboost for java interface

```
$ cd chapter8
$ cp -r ../chapter2/iris_classification_xgboost .
$ mv src/main.rs src/lib.rs
$
```

Now we will need to make some edits in the code. Notice that all the code was written in the run function, which was in turn called in the main function. We will not need the main function in this case and hence we will remove it. Then we will break the run function to two functions, fit and predict. The fit function will read the training file, arrange it according to relevant vectors, and then train the model. Once the model is trained, we will save the model in an xgb.model file (Listing 8-17).

Listing 8-17. chapter8/iris_classification_xgboost/iris_ classification_library/src/lib.rs

```
pub fn fit() -> Result<(), Box<dyn Error>> {
  let training_file = "data/iris.csv";
  let file = File::open(training_file).unwrap();
  let mut rdr = csv::Reader::from_reader(file);
  let mut data = Vec::new();
  for result in rdr.deserialize() {
    let r: Flower = result.unwrap();
    data.push(r); // data contains all the records
  }

  // previous code that was part of run function.

  // train model, and print evaluation data
  let booster = Booster::train(&params).unwrap();

  // save and load model file
  println!("\nSaving Booster model...");
  booster.save("xgb.model").unwrap();

  Ok(())
}
```

Similarly, we have the predict method, which skips out all the code for training and only has the data reading and organising code and the predict code (Listing 8-18).

Listing 8-18. chapter8/iris_classification_xgboost/iris_classification_library/src/lib.rs

```
pub fn predict() -> Result<String, Box<dyn Error>> {
  println!("Loading model");
  let booster = Booster::load("xgb.model").unwrap();
  let predict_file = "data/predict.csv";
  let file = File::open(predict_file).unwrap();
  let mut rdr = csv::Reader::from_reader(file);
  let mut data = Vec::new();
  for result in rdr.deserialize() {
    let r: Flower = result.unwrap();
    data.push(r); // data contains all the records
  }
  let val_size: usize = data.len();

  // differentiate the features and the labels.
  let flower_x_val: Vec<f32> = data.iter().flat_map(
    |r| r.into_feature_vector()).collect();
  let flower_y_val: Vec<f32> = data.iter().map(
    |r| r.into_labels()).collect();

  // validation matrix with 1 row
  let mut dval = DMatrix::from_dense(
    &flower_x_val, val_size).unwrap();
  dval.set_labels(&flower_y_val).unwrap();

  let preds = booster.predict(&dval).unwrap();
  Ok(flower_decoder(preds[0]))
}
```

Check that in the fit function, we train the model and save the model. Similarly, in the predict method, we load the model to perform the predictions. As you can probably understand, this method of persisting models on disk between runs is not ideal. Also, for each predict request, this will read the whole model again and again. A better way would be to persist the model in memory and then pass the pointers to the model, but this may be unstable and may have memory leaks in them. You can see that the same method has been seen in the Python section as well.

Java class Now that we have refactored the Rust class to cover the functionality, we will come back to the Java class definition. The main point to note is that we will need to define a class with the method names as same as the functions that we will call so that the jni is able to link the functions in an effective manner. So we define the signature like that shown in Listing 8-19.

Listing 8-19. chapter8/iris_classification_xgboost/iris_classification_library/src/lib.rs

```
class IrisClassificationXgboost {
  private static native void fit();
  private static native String predict();

  static {
    System.loadLibrary("iris_classification_xgboost");
  }
}
```

In the Java class in Listing 6-19, we have the class IrisClassificationXgboost and then we define native methods under it, fit and predict. The static `System.loadLibrary` will load the library and understand the functions and act as the appropriate linker.

Wrapper functions: Now we will write the appropriate linker functions that wrap over our functionality. To make things simpler, let's create a folder iris_classification_library and move the Rust code inside the folder. We will also need to make a few changes in Cargo.toml. Apart from the previous dependencies, we will need to add the jni dependency and we will need to specify the crate to be of type cdylib (Listing 8-20).

Listing 8-20. chapter8/iris_classification_xgboost/iris_classification_library/Cargo.toml

```
[package]
name = "iris_classification_xgboost"
version = "0.1.0"
edition = "2018"

[dependencies]
# previous dependencies
ml-utils = { path = "../../../chapter2/ml-utils" } # give
correct path to mlutils
jni = "0.12.3"

[lib]
name = "iris_classification_xgboost"
crate-type = ["cdylib"]
```

Now if we run cargo build in the directory, we should see a .dylib file or an .so (depending on your OS) file in target/debug folder. Now we need to define our exported methods (Listing 8-21).

Listing 8-21. chapter8/iris_classification_xgboost/iris_classification_library/src/lib.rs

```
use jni;
use jni::JNIEnv;
use jni::objects::{GlobalRef, JClass, JObject, JString};
use jni::sys::{jint, jlong, jstring, jbyteArray};
```

```
#[no_mangle]
#[allow(non_snake_case)]
pub unsafe extern "system" fn Java_IrisClassificationXgboost_
fit(_env: JNIEnv, _class: JClass) {
  fit().unwrap();
}


#[no_mangle]
#[allow(non_snake_case)]
pub unsafe extern "system" fn Java_IrisClassificationXgboost_
predict(
  env: JNIEnv,
  _class: JClass,
) -> jstring {
  let output = env.new_string(predict().unwrap())
      .expect("Couldn't create java string!");
  output.into_inner()
}
```

Notice in Listing 8-21 the dependencies that we are loading. JNIEnv is the interface to the JVM that the majority of the methods will work on. The objects in the jni::objects are the objects that carry additional lifetime information that prevent them from escaping the context. The objects in jni::sys are meant to return pointers to the appropriate data from Rust. We cannot send the exact data because the lifetime checker won't let us.

Now let's focus on the wrapper functions Java_IrisClassification Xgboost_fit and Java_IrisClassificationXgboost_predict. We will apply the no_mangle outer attribute to these wrapper functions, so that Rust does not mangle the method names when creating the binaries and Java is able to identify the functions. Generally, if this is not done when creating the binaries, Rust internally mangles the names of the methods

that are written in the code.[3] The Java needs to be written in the format
`Java_classname_methodname`. In this way, the `Java_IrisClassification`
`Xgboost_fit` and `Java_IrisClassificationXgboost_predict` wrapper
functions have been created.

Now the fit method is fine as this is a void method, but in case of
`Java_IrisClassificationXgboost_predict,` we are returning a string. In
this case we need to take care of some additional things. The output type
needs to be a jstring and hence we will need to create a new java string
using `env.new_string` from the predict function. We will then return the
pointer to the function.

Now we have written all the code that would expose the Rust functions
to Java code. We should now be able to run this code. For this we create the
main method in Java (Listing 8-22).

Listing 8-22. chapter8/iris_classification_xgboost/iris_
classification_library/src/lib.rs

```
class IrisClassificationXgboost {
  // previous code

  public static void main(String[] args) {
    IrisClassificationXgboost.fit();
    String predictions = IrisClassificationXgboost.predict();
    System.out.println(predictions);
  }
}
```

---

[3]Name mangling is used in various other languages such as FORTRAN as well
as to solve various problems caused by the need to solve unique names for
programming entities.

Notice in Listing 8-22 that we are retrieving the output of IrisClassificationXgboost.predict as a java string and then we are printing it. We should not be able to compile this code and run the functions (Listing 8-23).

Listing 8-23. chapter8/iris_classification_xgboost/iris_classification_library/src/lib.rs

```
$ cd iris_classification_library \
        && cargo build
$ javac IrisClassificationXgboost.java \
        && java -Djava.library.path=iris_classification_
        library/target/debug/ \
        IrisClassificationXgboost
[15:27:22] DANGER AHEAD: You have manually specified `updater`
parameter. The `tree_method` parameter will be ignored.
Incorrect sequence of updaters will produce undefined behavior.
For common uses, we recommend using `tree_method` parameter
instead.
[15:27:22] src/tree/updater_prune.cc:74: tree pruning end,
        1 roots, 2 extra nodes, 0 pruned nodes, max_depth=1
[15:27:22] src/tree/updater_prune.cc:74: tree pruning end,
        1 roots, 4 extra nodes, 0 pruned nodes, max_depth=2
[15:27:22] src/tree/updater_prune.cc:74: tree pruning end,
        1 roots, 4 extra nodes, 0 pruned nodes, max_depth=2
[0]     test-merror:0.066667    train-merror:0.044444
[15:27:22] src/tree/updater_prune.cc:74: tree pruning end,
        1 roots, 2 extra nodes, 0 pruned nodes, max_depth=1
[15:27:22] src/tree/updater_prune.cc:74: tree pruning end,
        1 roots, 4 extra nodes, 0 pruned nodes, max_depth=2
```

```
[15:27:22] src/tree/updater_prune.cc:74: tree pruning end,
           1 roots, 4 extra nodes, 0 pruned nodes, max_depth=2
[1]       test-merror:0.066667     train-merror:0.033333

Saving Booster model...
Loading model
setosa
```

The Java code was able to take the predictions and print it out. We can now encode these commands in a makefile (Listing 8-24).

Listing 8-24. chapter8/iris_classification_xgboost/Makefile

```
java_run: lib
        javac IrisClassificationXgboost.java \
            && java -Djava.library.path=iris_classification_
            library/target/debug/ \
            IrisClassificationXgboost

.PHONY: lib

javah:
        javah IrisClassificationXgboost

lib:
        cd iris_classification_library \
            && cargo build
```

In this way we should be able to create functions in Rust and call those applications in Java.

8.2 Rust in the Cloud

There are various benefits of using machine learning in the cloud.

- The cloud's pay-per-use model is good for bursty and erratic AI or machine learning workloads.

- The cloud makes it easy for enterprises to experiment with machine learning capabilities and scale up as projects go into production and demand increases.

- The barriers to building a machine learning application that provide value to the business are many: from expertise revolving around data sifting, building and training good machine learning models, to specialized hardware requirements for machine learning models. Good cloud providers focus on providing solutions to all these aspects.

In this section we will take a look at AWS Lambda. After designing and learning an ML model, the hardest part is actually running and maintaining it in production. Integrating serverless applications in the machine learning workflow allow for quick scaling of the machine learning application.

When creating serverless applications, we would generally select the runtimes to be one of Python or Node. Luckily, we have the aws runtime for Rust[4] and hence we are able to create serverless applications in Rust. The API's for AWS serverless applications define an HTTP-based specification of the Lambda programming model, which, ideally, can be implemented in any model.

In our application, we will be making a simple word count application. To create the application, we will need to add the dependencies to the

---

[4]https://github.com/awslabs/aws-lambda-rust-runtime.

Cargo.toml file. The serde dependency will serialize and deserialize the incoming data. We can use log and simple-logger for logging and regex to be able to parse through the incoming text. Finally, the most important thing is that we will need to add lambda_runtime to be able to create the serverless application (Listing 8-25).

Listing 8-25. chapter8/my_lambda_function/Cargo.toml

```toml
[package]
name = "my_lambda_function"
version = "0.1.0"
edition = "2018"

[dependencies]
lambda_runtime = "0.1"
serde = "^1"
serde_derive = "^1"
serde_json = "^1"
log = "0.4"
simple_logger = "^1"
regex = "1"
```

To be able to serialize and deserialize the incoming data, we will need to create an event struct and have serialization and deserialization added so that the runtime is able to parse the data (Listing 8-26).

Listing 8-26. chapter8/my_lambda_function/src/main.rs

```rust
use serde_derive;
use serde_derive::{Serialize, Deserialize};

#[derive(Serialize, Deserialize)]
struct CustomEvent {
  string: String,
}
```

In our lambda application, we will need to execute the handler as part of the main function. In the Listing 8-27, my_handler is the handler function that will have the handler code for the output on the event.

Listing 8-27. chapter8/my_lambda_function/src/main.rs

```
use lambda_runtime;
use lambda_runtime::{lambda, Context, error::HandlerError};
use log;
use log::error;
use std::error::Error;

fn main() -> Result<(), Box<dyn Error>> {
  simple_logger::init_with_level(log::Level::Debug).unwrap();
  lambda!(my_handler);

  Ok(())
}
```

We should now be able to put in the code for the handler. The incoming data will be parsed and put in the CustomEvent struct using serde. If the string is not given, then we return saying that no input string has been provided. If a string is present, then we create a hashmap with all the words in the string, with the values as the count of the words in the string. The words in the string are recognized by parsing using the regex module. Last, we will need to serialize to a string, which we perform using json serialization (Listing 8-28).

Listing 8-28. chapter8/my_lambda_function/src/main.rs

```
use std::collections;
use std::collections::hash_map::Entry::{Occupied, Vacant};
use regex;
use regex::Regex;
```

```rust
fn my_handler(
  event: CustomEvent, ctx: Context)
  -> Result<String, HandlerError> {
  if event.string == "" {
    error!("Empty string in request {}", ctx.aws_request_id);
    return Err(ctx.new_error("Empty input string"));
  }
  let mut map = collections::HashMap::<String, u32>::new();
  let re = Regex::new(r"\w+").unwrap();
  for caps in re.captures_iter(&event.string) {
    if let Some(cap) = caps.get(0) {
      let word = cap.as_str();
      match map.entry(word.to_string()) {
        Occupied(mut view) => { *view.get_mut() += 1; }
        Vacant(view) => { view.insert(1); }
      }
    }
  }
  let j = serde_json::to_string(&map).unwrap();
  Ok(j)
}
```

Now although the app by itself is done, we would need to perform some additional hoops to be able to run it in the lambda runtime on the cloud servers. When configured to use a custom runtime with the Runtime APIs, AWS Lambda expects the deployment package to contain an executable file called bootstrap. We can configure Cargo to generate a file called bootstrap, regardless of the name of our crate. First, in the [package] section of the file, add an autobins = false setting. Then, at the bottom of the Cargo.toml, add a new [[bin]] section providing the name of the binary file and the path of the code (Listing 8-29).

Listing 8-29. chapter8/my_lambda_function/Cargo.toml

```
[package]
name = "my_lambda_function"
version = "0.1.0"
authors = ["joydeep bhattacharjee"]
autobins = false
edition = "2018"

# dependencies code

[[bin]]
name = "bootstrap"
path = "src/main.rs"
```

This would create a binary executable name bootstrap when built in line with the AWS expectations.

Now before we build, we need to make sure that the Rust compiler is targeting the correct platform. AWS Lambda executes in an Amazon Linux environment and hence we will need to build against this specific platform. This is where Cargo as a tool shines again as it makes it very easy for us to be able to build cross-platform tools. So, we don't necessarily need to compile in an Amazon Linux machine but can build on, let's say, a MacOS laptop as well. First we will need to install the musl tool to be able to build against the platform (Listing 8-30).

Listing 8-30. rustup add linux target

```
$ rustup target add x86_64-unknown-linux-musl
$
```

Before we build the platform, we will need to add a linker for the compilation toolchain. This means that we will need to run a brew install command on MacOS platforms.[5] See Listing 8-31.

Listing 8-31. Brew install musl-cross

```
$ brew install filosottile/musl-cross/musl-cross
$
```

Once done, we will need to tell Cargo that when building, use this linker instead the default one. This is done by creating a new file in .cargo/config with the linker information (Listing 8-32).

Listing 8-32. chapter8/my_lambda_function/.cargo/config

```
[target.x86_64-unknown-linux-musl]
linker = "x86_64-linux-musl-gcc"
```

Sometimes the linker would not get picked up directly, so we might need to manually soft link to the linker (Listing 8-33).

Listing 8-33. link musl gcc

```
$ ln -s /usr/local/bin/x86_64-linux-musl-gcc /usr/local/bin/
musl-gcc
$
```

Now with the toolchain fully in place, we should be able to create the executable that can be run on the lambda platform (Listing 8-34).

Listing 8-34. cargo build

```
$ cargo build --release --target x86_64-unknown-linux-musl
$
```

---

[5]https://github.com/FiloSottile/homebrew-musl-cross.

We should now be able to see the executable built in the target folder (Listing 8-35).

Listing 8-35. check executable fi le

```
$ ls target/x86_64-unknown-linux-musl/release/
bootstrap       bootstrap.d     build       deps
examples        incremental     native      runtime_release
$
```

We will need to zip this executable to be able to deploy to AWS Lambda (Listing 8-36).

Listing 8-36. zip for AWS

```
$ zip -j rust.zip ./target/x86_64-unknown-linux-musl/release/
bootstrap
$
```

Since this involves quite a lot of commands, we can put them in a simple build shell file (Listing 8-37).

Listing 8-37. chapter8/my_lambda_function/buildthis.sh

```
rm -f rust.zip
cargo build --release --target x86_64-unknown-linux-musl
zip -j rust.zip target/x86_64-unknown-linux-musl/release/
bootstrap
```

We can now deploy this file to AWS Lambda. Navigate to the AWS Lambda console and create a new function. We will need to select Author from Scratch. Type in a name of the function and, importantly, select the runtime to be "Provide your own bootstrap." The page should look something like Figure 8-1.

In the next page, scroll below to function code. Click on code entry type and then select "Upload a .zip file." See Figure 8-2

Now we need to select the correct file and click the save button. This is shown in Figure 8-3.

Once saved, our lambda function is created. We can now configure a test case in the console to see if the function is behaving as expected. This is shown in Figure 8-4. Since our function expects a string parameter, we can pass a random sentence to the string parameter, and that should give us the word count.

In Figure 8-5 we see that running the test case gives us the correct result.

We have been successful in creating a serverless application in Rust. There are some caveats in this model of bprogramming though. In our case, the algorithm was simple and can be built easily using Rust. If the code is in pure Rust, the Rust tooling that is built works, and we should be able to build serverless applications easily. In case the applications have external library dependencies, such as the ones we have seen in the majority of this book, it is quite difficult to perform cross-linking and compilation and create a binary for the AWS platform. Still if we are able to do that, then model inference will work. Model training using serverless Rust is not shown because that takes up time and serverless applications have a timeout. Nonetheless, integrating serverless applications in the ML workflow gives a huge productivity benefit to the developer.

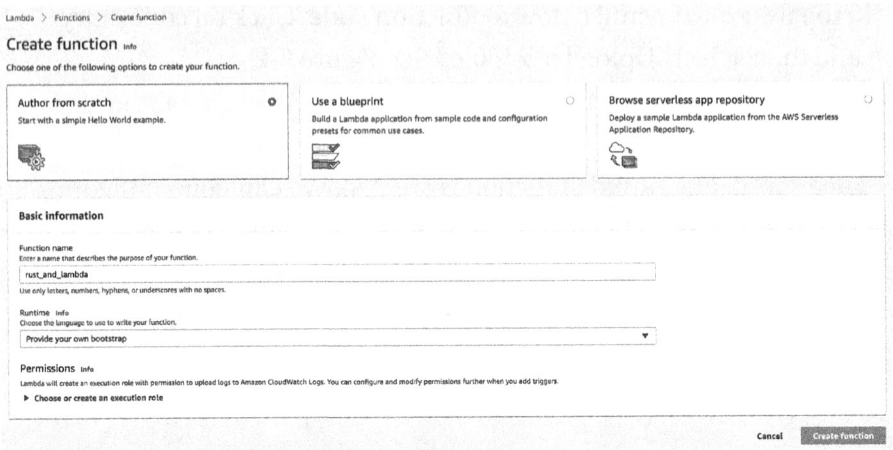

Lambda > Functions > Create function

Create function Info

Choose one of the following options to create your function.

Author from scratch ⊙	Use a blueprint ○	Browse serverless app repository ○
Start with a simple Hello World example.	Build a Lambda application from sample code and configuration presets for common use cases.	Deploy a sample Lambda application from the AWS Serverless Application Repository.

Basic information

Function name
Enter a name that describes the purpose of your function.

    rust_and_lambda

Use only letters, numbers, hyphens, or underscores with no spaces.

Runtime Info
Choose the language to use to write your function.

    Provide your own bootstrap   ▼

Permissions Info

Lambda will create an execution role with permission to upload logs to Amazon CloudWatch Logs. You can configure and modify permissions further when you add triggers.

▶ Choose or create an execution role

Cancel   **Create function**

Figure 8-1. *Create lambda funcion*

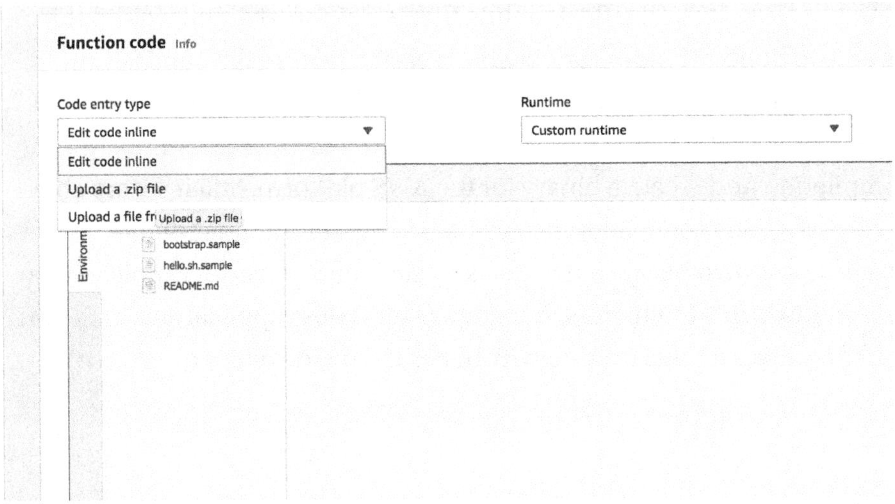

Function code Info

Code entry type   Runtime

    Edit code inline   ▼   Custom runtime   ▼

    Edit code inline

    Upload a .zip file

    Upload a file fr Upload a .zip file

    [icon] bootstrap.sample
    [icon] hello.sh.sample
    [icon] README.md

Figure 8-2. *Select upload method*

344

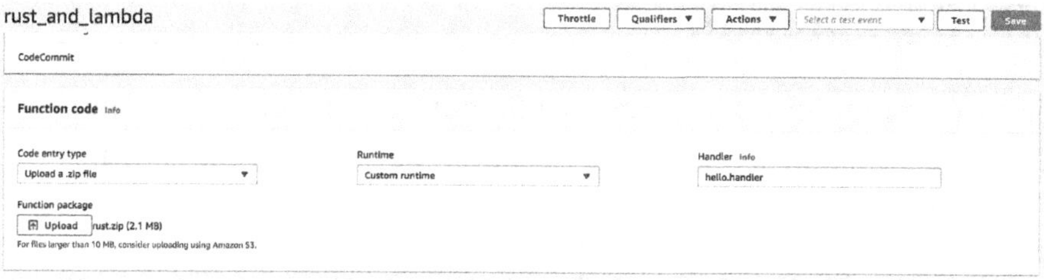

rust_and_lambda   Throttle | Qualifiers ▼ | Actions ▼ | Select a test event ▼ | Test | Save

CodeCommit

Function code Info

Code entry type
Upload a .zip file ▼

Runtime
Custom runtime ▼

Handler Info
hello.handler

Function package
⊞ Upload rust.zip (2.1 MB)
For files larger than 10 MB, consider uploading using Amazon S3.

Figure 8-3. *Select zip file*

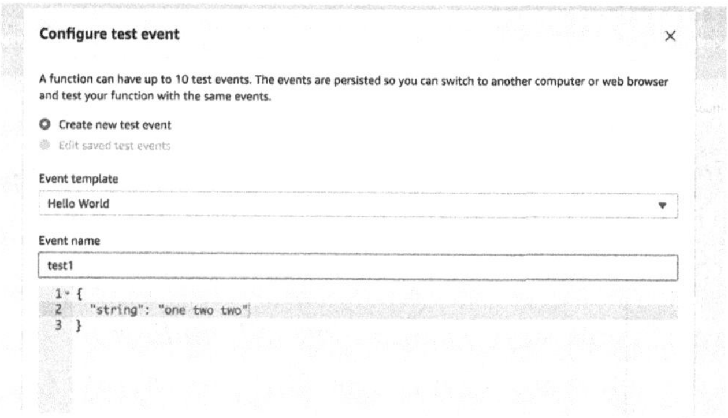

Configure test event ✕

A function can have up to 10 test events. The events are persisted so you can switch to another computer or web browser and test your function with the same events.

◉ Create new test event
○ Edit saved test events

Event template
Hello World ▼

Event name
test1

```
1 ▾ {
2    "string": "one two two"
3 }
```

Figure 8-4. *Test case*

rust_and_lambda

⊘ Execution result: succeeded (logs)

▾ Details

The area below shows the result returned by your custom runtime function executic

"{\"one\":1,\"two\":2}"

Summary

Code SHA-256

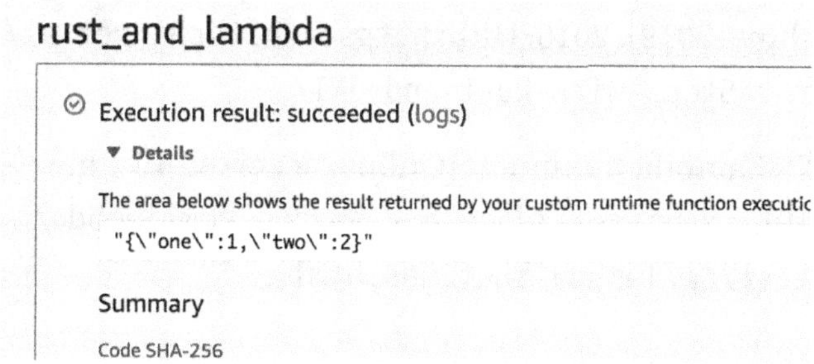

Figure 8-5. *Test run*

8.3 Conclusion

This last chapter introduced you to different ways in which Rust code can be integrated in an existing application. The chapter starts with writing wrappers for Python code using PyO3, and then we moved on to writing JNI wrapper code for using Rust code in Java. Finally, we created a serverless application in AWS Lambda for an example of Rust in the cloud.

8.4 Bibliography

[1]   Ben Frederickson. *Writing Python Extensions in Rust Using PyO3*. "[On- line; accessed 5-June-2019]". 2018. url: `https://www.benfrederickson.com/writing-python-extensions-in-rust-using-pyo3/`.

[2]   Carol Nichols Steve Klabnik. *Safe JNI Bindings in Rust*. [Online; accessed 19-Nov-2019]. url: `https://docs.rs/jni/0.12.3/jni/`.

[3]   Michael Nitschinger. *First Steps with Rust and JNI*. Ed. by Machine Learn-ing Mastery. [Online; accessed 11-June-2019]. 2016. URL: `https://nitschinger.at/First-Steps-with-Rust-and-JNI/`.

[4]   AWS Lambda Runtimes. [Online; accessed 15-June-2019]. url: `https://docs.aws.amazon.com/lambda/latest/dg/lambda-runtimes.html`.

Index

© Joydeep Bhattacharjee 2020
J. Bhattacharjee, *Practical Machine Learning with Rust*,
https://doi.org/10.1007/978-1-4842-5121-8

Printed in the United States
By Bookmasters